CHEMISTRY AND PHYSICS OF CARBON

Volume 19

CHEMISTRY AND PHYSICS
OF CARBON

A SERIES OF ADVANCES

Edited by
Peter A. Thrower

DEPARTMENT OF MATERIALS SCIENCE AND ENGINEERING
THE PENNSYLVANIA STATE UNIVERSITY
UNIVERSITY PARK, PENNSYLVANIA

Volume 19

CRC Press
Taylor & Francis Group
Boca Raton London New York

CRC Press is an imprint of the
Taylor & Francis Group, an **informa** business

First published 1984 by Marcel Dekker, Inc.

Published 2021 by CRC Press
Taylor & Francis Group
6000 Broken Sound Parkway NW, Suite 300
Boca Raton, FL 33487-2742

© 1984 by Taylor & Francis Group, LLC
CRC Press is an imprint of Taylor & Francis Group, an Informa business

No claim to original U.S. Government works

ISBN 13: 978-0-8247-7245-1 (hbk)

**Visit the Taylor & Francis Web site at
http://www.taylorandfrancis.com**

**and the CRC Press Web site at
http://www.crcpress.com**

The Library of Congress Cataloged the
First Issue of This Title as Follows:

Chemistry and physics of carbon, v. 1-
    London, E. Arnold; New York, M. Dekker, 1965-

        v. illus. 24 cm

    Editor: v. 1- P. L. Walker

    1. Carbon. I. Walker, Philip L., 1928- ed.

QD181.C1C44   546.681

Library of Congress   1   66-58302
ISBN 0-8247-7245-8

Preface

This volume should provoke considerable interest as it contains contributions from Great Britain, Yugoslavia, the U.S.A., and the USSR. It bears testimony to the internationality as well as the variety of carbon research.

It has long been recognized that impurities play tremendously important roles in the physical and chemical properties of carbons and graphites. They have been shown to catalyze oxidation and graphitization and can have significant effects on mechanical and electromagnetic properties. In the last few years, the introduction of species between the graphite layer planes to produce intercalation compounds has been extensively studied. However, in many instances the location and distribution of the impurity atoms in the graphite are uncertain. In the first chapter of this volume, Dr. Marinković reviews the evidence for some elements substituting directly for carbon atoms in the graphite lattice. The ability of boron to do this has long been recognized and now evidence that silicon and phosphorous also have this ability is presented. This useful review should be of interest to all involved with graphite materials and the way their properties are influenced by impurities.

The second chapter of this volume is concerned with a subject which receives coverage in previous volumes of this series. In the very first volume, Palmer and Cullis discuss "The Formation of Carbons from Gases," and in Volume 5, Bokros reviews the "Deposition, Structure and Properties of Pyrolytic Carbon." Volume 9 contains a chapter on the "Deposition of Pyrolytic Carbon in Porous Solids" by Kotlensky, and several other contributors discuss the formation of carbons from gases to some extent. The work of Dr. Tesner, of the

All-Union Research Institute of Natural Gas, in Moscow, USSR, is
often referred to by these earlier contributors, and it is indeed
a pleasure to be able to include his chapter on the "Kinetics of
Pyrolytic Carbon Formation" in the present volume. In this contribution, he discusses the formation kinetics of carbon from various
gases on a variety of substrates. The use of a binary mixture of
hydrocarbons is treated, as is the inhibiting effect of hydrogen.
Workers in many areas of carbon research should find this contribution both interesting and informative.

The substitutional nature of boron atoms in graphite was used
by Hennig in an elegant experiment to determine the diffusion coefficient of boron atoms in the graphite lattice. This was accomplished
using the etch decoration technique, which was pioneered by Hennig
in the 1960s; many of the results of this technique are reviewed in
Volume 2 of this series. During the 1970s, several people attempted,
unsuccessfully, to use the etch decoration technique, and only recently has a detailed procedure been described which gives reproducible results. The work leading to this procedure was performed under
the direction of Dr. Yang, who contributes the third chapter in this
volume. As a result, researchers involved with gas-carbon reactions
now have a tool available that enables them to explore these phenomena
on a truly microscopic scale.

The optical properties of carbon are the subject of a chapter by
Sabri Ergun in Volume 3 of this series. In the last chapter of the
present volume, a theory is presented by R. A. Forrest et al. which
can be used to interpret the results of optical microscopy studies
on anisotropic carbons. The value of their contribution lies in its
applicability to those carbonaceous materials which are of everyday
interest. In recent years, there has been a tremendous interest in
carbonization phenomena, with special interest in optical microscopy
studies. Researchers in this area, both industrial and academic,
should find this contribution of particular interest.

I am confident that there is something in this volume to interest
all carbon researchers and trust that it will serve to provoke research
and even clearer insight into some of the problems discussed.

Peter A. Thrower

Contributors to Volume 19

*C. Cornford** Northern Carbon Research Laboratories, University of Newcastle upon Tyne, Newcastle upon Tyne, England

R. A. Forrest[†] Northern Carbon Research Laboratories, University of Newcastle upon Tyne, Newcastle upon Tyne, England

B. T. Kelly Springfields Nuclear Power Development Laboratories, United Kingdom Atomic Energy Authority (Northern Division), Springfields, Salwick, Preston, England

S. Marinković Department of Materials Science, Boris Kidrič Institute of Nuclear Sciences-Vinča, Belgrade, Yugoslavia

H. Marsh Northern Carbon Research Laboratories, University of Newcastle upon Tyne, Newcastle upon Tyne, England

P. A. Tesner All-Union Research Institute of Natural Gas, Moscow, USSR

Ralph T. Yang Department of Chemical Engineering, State University of New York at Buffalo, Amherst, New York

Present affiliation: Integrated Geochemical Interpretation Ltd., Hallsannery, Bideford, Devon, England
†*Also affiliated with:* U.K.A.E.A., Harwell, Oxfordshire, England

Contents of Volume 19

Contents of Other Volumes

CHEMISTRY AND PHYSICS
OF CARBON

Volume 19

1

Substitutional Solid Solubility in Carbon and Graphite

S. MARINKOVIĆ

Department of Materials Science
Boris Kidrič Institute of Nuclear Sciences—Vinča
Belgrade, Yugoslavia

I. INTRODUCTION

The scope of a discussion of substitutional solid solubility in
graphite and carbons is susceptible to somewhat different ideas.
In the strict sense of the term, substitutional solid solubility
means statistical substitution of the carbon atoms in the lattice

by some atoms other than carbon to obtain a thermodynamically stable homogeneous solid solution.

Graphite and carbons are produced by pyrolysis of organic substances which in addition to hydrogen may contain O, N, S, or other heteroatoms which can remain in the resulting carbon structure up to high temperatures [1,2]. It is also possible to introduce such elements into carbons by chemical means [2]. However, the question remains whether these foreign atoms are substituted for carbon atoms in the carbon lattice or are just chemically bound to the macromolecules. Marchand, who reviewed the literature data on doping of carbon [2], concluded that only boron was known with certainty to be located at trigonal sites of the carbon/graphite lattice. We have decided that it would be more justifiable to leave these elements out of this review.

Another question that may be posed is whether a review of substitutional solid solubility should deal with what is meant by catalytic graphitization. The term is applied to graphitization favored by an addition of foreign elements or inorganic substances. It has often been used to designate increased graphitization by an addition of a higher percentage of a catalyst (5 to 10 percent) where mechanisms other than doping action are operative. However, the term encompasses different mechanisms leading to graphitization, including doping action [3-5].

In this chapter we deal only with doping, and therefore articles dealing with catalytic graphitization will be considered only to the extent with which they deal with doping. Results concerning doping activity have already been reviewed by other authors. Articles in this series by Fischbach [3] and Marchand [2] have dealt with the catalytic graphitization and electronic properties of doped carbons, respectively. There are also the recent articles by Ōya [4,5] concerning catalytic graphitization, including doping.

The question may therefore be raised as to the need for a review article on the subject of substitutional solid solubility in carbon/ graphite. The author can see at least several reasons in favor of such a review: (1) there is new knowledge recently gathered on the subject; (2) the reviews mentioned treated broader topics and included doping only to the extent, or from the point of view, compatible with the subject treated; and (3) the subject is sufficiently interesting from both theoretical and practical aspects to deserve attention and some stimulation of interest.

The experimental evidence presented is arranged according to the elements whose solid solubility is being considered. Section II, which deals with boron, is presented in order of decreasing graphitization level of the matrix carbon material.

Section III, about silicon, where the available experimental evidence is restricted to medium temperature pyrolytic carbon, is subdivided into two parts, treating structural and electronic properties. Conclusions about the substitutional solid solubility of silicon in carbon are derived on the basis of the experimental facts presented.

Section IV deals with the relatively modest experimental evidence relative to the carbon-phosphorus system. Although it is perhaps too early to draw conclusions as to the solid solubility of phosphorus in carbon, the author feels that the data presented may be useful at least in stimulating further research on this point.

II. BORON

The boron atom has three electrons in the outer shell, two s electrons and one p electron. However, B is always trivalent, because the energy per bond liberated by the formation of three bonds is much higher than the energy of formation of a single bond. This energy allows the excitation of the B atom into a hybridized sp^2 state. The three sp^2 orbitals are coplanar with angles of 120° between them.

Although for the calculated radius of the trigonal B atom values of 0.085 to 0.090 nm are obtained [6], the experimental values are considerably (by at least 0.012 nm) smaller and thus very close to the radius of the C atom in graphite. Due to the incomplete electron octet of B in the trivalent B compounds, it has acceptor properties.

Thus the B atom can be regarded as ideal for substitution for a carbon atom in graphite: with its three electrons, it forms three coplanar orbitals, exactly as the carbon atom in graphite. The similarity of the atomic volumes should not significantly change the geometrical parameters of the lattice and, consequently, interactions between the lattice and charge carriers.

The interactions of a small amount of B with graphite and carbons has been studied extensively by many authors for a number of years. As early as 1894 a French patent claimed that an addition of 2 percent B_2O_3 to carbon had a favorable effect on graphitization. In order to systematize the large amount of knowledge available on this subject, it is presented in the order of decreasing graphitization level of the matrix material.

A. Graphite

Turnbull et al. [7] have prepared graphite single crystals containing up to 4 percent B by heating the amorphous B in a reactor-grade graphite crucible to about 2800 K. The lattice parameters of the as-prepared crystals, which were found to contain between 2 and 4 percent B, were a = 246.81 to 246.87 pm and c = 671.58 to 673.21 pm (compared to a = 246.12 pm and c = 670.90 pm for the undoped, well-annealed Ticonderoga natural graphite). No systematic relation between the interlayer spacing and B content was found.

Heat treatment of the boronated graphite resulted in a change of the lattice parameters (Fig. 1): c began to fall at about 1470 K, reached a minimum value which lay below that of the undoped natural graphite, and above 2100 K gradually increased toward the pure graphite value. The B content remained stable up to 2100 K and gradually decreased at higher heat-treatment temperatures (HTTs).

The authors suppose that the boronated graphite crystals contain B both in substitutional and interstitial positions. The substitutional B atoms lead to a contraction of c and an increase of a. The presence of the interstitial B atoms leads to an increase of c, the effect being greater than the contraction due to the substitutional B atoms. The presence of the interstitial atoms does not influence a.

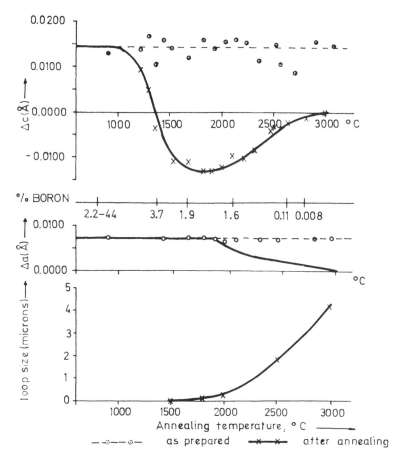

FIG. 1 Change in graphite lattice parameters (Δa and Δc) and loop size as a function of annealing temperature. Boron content corresponding to a given temperature is indicated. (From Ref. 7.)

By annealing up to about 1700 K, the interstitial B diffuses out of the crystal. The substitutional B atoms remain in their positions up to about 2300 K, when by a self-diffusion process they diffuse out, leaving a supersaturation of vacancies which precipitate out as vacancy loops visible by transmission electron microscopy (TEM). From the fact that the B content decreases more rapidly if the annealed samples are in the form of foils and from the behavior of the loops on annealing, it follows that B can also diffuse along the c axis.

The energy controlling the annealing behavior of the loops was found to be equal to the energy of self-diffusion. This is in accord with the statement that the loops are purely vacancy loops and not plates of precipitated substitutional B, because the energy corresponding to the latter case would be characteristic of B diffusion in graphite and different from the energy of self-diffusion.

Also, bearing in mind the known interatomic B-B distance in other materials (about 0.165 nm), it does not seem probable that a precipitate of the observed dimensions could conform with the precise geometrical arrangement imposed by the graphite lattice, such as implied from the loop observations.

A more precise determination of the B solubility in graphite was made by Lowell [8] for the temperature range 2070 to 2770 K (Fig. 2). According to his results, the solid solubility limit of B increases from 1.00 atomic percent at 2070 K to 2.35 atomic percent

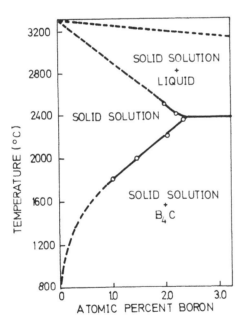

FIG. 2 Part of the B-C phase diagram showing the substitutional solid solubility of B in graphite. (From Ref. 8.)

at 2620 K. At higher temperatures the solubility decreases. The lattice parameters were found to be linear functions of B content:

$$a = 246.023 + 0.310K_B$$
$$c = 671.163 - 0.594K_B$$

Here K_B is B concentration in atomic percent, and a and c are lattice parameters in pm.

Since c decreases and a increases with B content, it is supposed that B atoms are located in the graphite layers rather than between them. Two sites were considered possible for the B atoms: (1) at the center of a hexagon of C atoms, and (2) in a substitutional position. By comparing the helium density with density values calculated for these two hypotheses, good agreement was found for hypothesis 2—the substitutional solid solution.

Wagner and Dickinson [9] have measured the physical properties of B-doped, resin-bonded polycrystalline graphite as a function of B content and temperature. In the dopant concentration range investigated (up to 0.79 wt percent), a single phase product is obtained. The interlayer spacing decreases with B content, but the crystallite size perpendicular to the layers (L_c) and the preferred orientation remain virtually unaffected. The coefficient of thermal expansion decreases with increasing B content, which is a logical consequence of the decrease in the interlayer spacing. The Young's modulus values increase with B content.

Electrical resistivity measured at room temperature decreases with B content. However at higher temperatures (above room temperature) the electrical resistivity first increases with B content, then passes through a maximum, and finally decreases. The authors conclude that it is difficult to reconcile the effects of B doping and temperature on electrical resistivity with the concept that the only effect of B is to lower the Fermi level.

Thermal conductivity (λ) was found to decrease with B content, which is a result that disagrees with that of Bowman et al. [10], the effect being distinct at low temperature, where a plot of λ as a function of temperature passes through a maximum. The results, in

particular the observed c contraction, the behavior of λ, and the
thermal stability of B, are more consistent with the picture of
substitutional B doping.

The effect of B on the electronic properties of graphite was
studied by Soule [11] and Delhaes and Marchand [13]. Soule [11]
investigated the Fermi level shift in a graphite single crystal by
following the diamagnetic susceptibility and change in concentration
of holes produced by low concentrations of B atoms (from ∿1 ppm to
0.5 percent). Pure B was introduced by diffusion at 3270 K for 1 h
in an Ar atmosphere. Beginning at about 0.01 percent B, the susceptibility decreases from -22.1×10^{-6} emu/g to -1.9×10^{-6} emu/g
(Fig. 3). Comparison of the experimental curve with the theoretical
one, calculated according to McClure's susceptibility theory [12],
shows an ionization efficiency of only 67 percent instead of the
expected 100 percent, from which the shift in the Fermi level was
calculated (Fig. 4). The Fermi level of graphite which lies 0.027 eV
above the bottom of the conduction band (due to the overlap of the

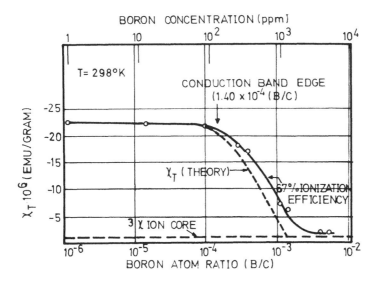

FIG. 3 Variation with B concentration of the total diamagentic
susceptibility χ_T at room temperature. Experimental values of Soule
[11] for B-doped single-crystal graphite; theoretical curve of McClure.
(From Ref. 12.)

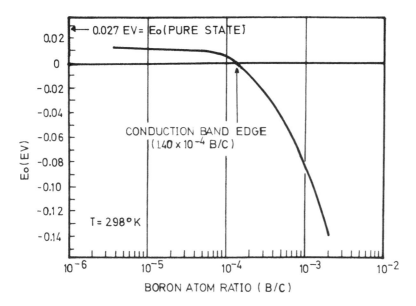

FIG. 4 Fermi level as a function of B content calculated from the
diamagnetic susceptibility of B-doped single-crystal graphite.
(From Ref. 11.)

valence and conduction bands) decreases, and at about 1.40×10^{-4} B/C
intersects the zero axis to enter, above this concentration, into the
valence band (i.e., the range of pure conductivity by holes). With
further doping the Fermi level continues to decrease rapidly to reach
a value of about -0.14 eV at about 2×10^{-3} B/C.

Results of Hall coefficient measurements confirm the conclusions
derived from the diamagnetic susceptibility (Fig. 5). A maximum in
Hall coefficient at room temperature due to the transition from mixed
electron/hole conduction (in the region of band overlap) to pure hole
conduction (in the valence band) was found at about the same B concentration ($\sim 2.2 \times 10^{-4}$ B/C) as that found from susceptibility measurements. This peak is shifted toward a lower B concentration of
3.3×10^{-5} B/C when the temperature is lowered to 77 K.

Electrical conductivity and magnetoresistance were found to
decrease with B level. The negative slope of the electrical conductivity is thought to be the result of an increased scattering by B

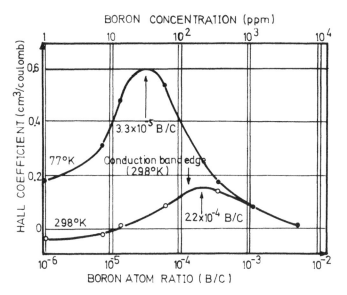

FIG. 5 Hall coefficient versus B content of the B-doped single-
crystal graphite. (From Ref. 11.)

atoms over that of the increased number of holes, contrary to the
results for polycry talline graphite. The reason probably lies in
the fact that the resulting mobility in a single crystal is of the
order of 10^5 cm^2 V^{-1} s^{-1} and is much more sensitive to the introduction of scattering centers than is the case for polycrystalline
graphite, where the presence of scattering centers in the original
material has already reduced the mobility to about 10^3 cm^2 V^{-1} s^{-1},
so that scattering due to a relatively small number of B atoms would
cause little difference.

Magnetoresistance decreases more rapidly because of the lack of
compensating influence of an increased number of holes. The mobility
of holes was found to depend on B concentration as (percent B)$^{-0.86}$.
An evaluation of the number of holes introduced by B doping gives an
efficiency of 75 ± 15 percent both from the susceptibility and Hall
coefficient measurements, which is less than could be expected for
such a perfect substitute as the B atom. A certain number of B
atoms, however, becomes trapped in various defects and these atoms
may be inefficient in the Fermi level shifting.

Polycrystalline graphite doped by B was investigated by Delhaes and Marchand [13], Grosewald and Walker [14], and Van der Hoeven et al. [15]. Delhaes and Marchand studied the electronic properties of B-doped polycrystalline graphite [13]. B was introduced by impregnation of graphite with boric acid followed by HT. The observed changes of diamagnetic susceptibility with the Fermi level position of boronated graphite are in agreement both with Soule's results for graphite single crystals [11] and with McClure's theoretical predictions [12]. Thus the diamagnetic susceptibility decreases with increasing B content, as Soule observed for graphite single crystals. The susceptibility versus 1/T curves show maxima for the samples with higher B content, in accordance with McClure's predictions (Fig. 6).

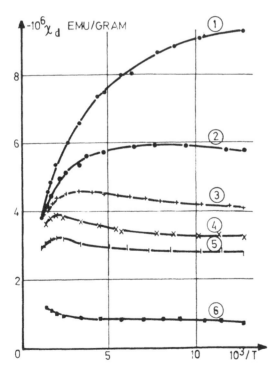

FIG. 6 Diamagnetic susceptibility versus temperature curves for polycrystalline graphite containing different B concentrations, increasing from ① to ⑥. (From Ref. 13.)

The Fermi level shift calculated from the Hall coefficient was
similar to the values obtained from diamagnetic susceptibility measurements. The efficiency of doping was found, however, to be of
the order of 10 percent, which is very little compared to Soule's
results for the graphite single crystal (65 to 75 percent). The
reason may lie either in the inadequacy of the experimental methods
used for B doping (possibly preventing some of the B atoms from
entering into substitutional lattice sites), or in a reduced diffusion due to the greater number of structural defects in polycrystalline graphite.

Preil et al. [16] have determined the dielectric function of
graphite doped with 0.5 atomic percent B. From measurements of
resistivity, reflectivity, and thermoreflectivity at low temperatures
(77 to 300 K) the authors have determined the Fermi level shift to be
0.27 eV. Using this value, an ionization efficiency of 41 to 42 percent has been determined, which is between the values given by Soule
[11] and Delhaes and Marchand [13].

The authors conclude that the properties investigated depend on
temperature and duration of HT in addition to their dependence on B
content. At longer times and higher HTT the evolution of the electronic properties is reversed, although the nominal B content remains
nearly unchanged. This phenomenon suggests that the HT of graphite
with boric acid first favors B diffusion and its substitution in the
lattice, but later leads to the formation of B_4C clusters.

Investigation of the thermoelectric power of samples with a B/C
ratio ranging from 10^{-6} to 10^{-2} [14] and of the low-temperature specific heat of boronated polycrystalline natural graphite with 0.23
atomic percent B [15] shows that the behavior of these properties is
consistent with expectations based on the supposition of B substitution in the graphite lattice.

Using a direct method of observation of B atoms by TEM, Hennig
[17] has determined the diffusion constants of B in graphite in the
temperature range 1970 to 2760 K. He made use of the observation
that lattice defects in graphite, such as vacancies or B atoms, as

well as the carbon atoms surrounding these defects, are more reactive
than the normal lattice constituents. This fact enabled the author
to make the B atoms in the cleaved graphite surface visible, using
the etch decoration method he had developed earlier. Diffusion was
provoked by heating the graphite crystals, with an established concentration gradient of B atoms, to different temperatures.

The mobility of B was found to be relatively insensitive to B
concentration for the B/C range between 10^{-2} and 10^{-8}. Although the
diffusion constants for the directions parallel ($6320e^{-157,000/RT}$)
and perpendicular ($7.1e^{-153,000/RT}$) to the graphite planes differ by
a factor 400 at 2070 K, the activation energies do not differ appreciably. The value of about 155 kcal/mole obtained for both directions is surprisingly close to the heat of sublimation of graphite
(172 kcal/mole) and to the activation energy for self-diffusion
(163 kcal/mole). However, the diffusion constants for self-diffusion
are less by at least 10^3 from those for B diffusion parallel to the
layers.

The results of Wagoner's ESR experiments on B-doped graphite
single crystals relative to the g shift as a function of B content
[18] could not be explained by McClure and Yafet on the basis of the
assumption that the only effect of substituting B for C was to add a
hole to the valence band. Another possible effect of B was discussed,
but no quantitative study of it has been made.

Evidence accumulated over years of investigations on boronated
graphite allows the conclusion to be made that a B atom can enter the
graphite lattice substituting for carbon atoms.

The evidence suggesting this conclusion comprises:

A considerable shift of the Fermi level as deduced from measure-
        ments of electronic properties (diamagnetism, galvanomag-
        netic properties)

The existence of linear relations between lattice parameters
        a (increase) and c (decrease) and B content

The absence of another phase containing B, up to a concentration
        supposed to be the limit of substitutional solid solubility
        and its sudden appearance above that limit

The behavior of boronated graphite on HT

A comparison of the density calculated for the hypothesis of
    substitutional solid solution, with experimental He density

The low B concentration region of the B-C phase diagram is proposed for a high temperature range between 2070 and 2520 K, according
to which the solid solubility of B increases linearly with temperature
from 1.00 to 2.35 atomic percent B.

The question that remains incompletely solved concerns a possibility of interstitial solid solubility of B in graphite at lower
temperatures (up to about 1700 K). In order to explain the lower
efficiency of doping than expected for B atoms, it is assumed that
a certain number of B atoms is trapped in various defects.

A number of properties of boronated graphite are influenced by
the presence of B in the lattice. Their dependence on B content can
be understood, at least qualitatively, on the basis of existing
knowledge.

B. Graphitizable Carbons

In their early work relative to pyrocarbon (PC) and boronated pyrocarbon (BPC) thin film resistors, Grisdale et al. [19] have found
that B considerably affects electrical resistivity and the temperature coefficient of resistivity of the films, both of these properties being substantially lowered by B addition. The minima in these
properties were found to appear at about 4 percent B. The effects
of B were ascribed principally to its graphitizing action, although
other effects, among them formation of cross-links between atomic
layers in adjacent crystal packets, and substitutional solid solubility of B in the graphite lattice, were also admitted.

In this case the catalytic graphitization effects of a higher
B content [4-5] are apparently superimposed on the effect of B doping,
both of which tend to increase electrical conductivity. There is also
an inverse effect of increasing B_4C content.

Albert and Parisot [20] have studied the effect of B on graphitization, and also its effect on the electrical and thermal conduc-
tivities of graphitizable carbons. The carbons were mixed with B_2O_3

and subsequently heat treated. The presence of B was found to induce graphitization. The same degree of graphitization was achieved with B at a HTT of 2770 K as it was without B at a HTT of 3250 K.

Measurements of electrical and thermal conductivity confirm the authors' hypothesis that B substitutes for C in the lattice. Namely, electrical conductivity is increased by the presence of B because of an increase in the number of holes as charge carriers, and the effect is reversed when the B is removed. On the other hand, thermal conductivity is decreased by the presence of B, which, presumably, hinders the lattice vibrations.

In the work of Kotlensky [21] boron-doped PC (BPC) was prepared at 2270 and 2370 K. The degree of graphitization, $\bar{g}$, as a function of B content showed a distinct maximum at about 1 percent B (Fig. 7). Qualitatively, a similar dependence but with considerably higher $\bar{g}$ values was found for BPC heat-treated at the deposition temperature for 20 h. The existence of a maximum in the degree of graphitization at a certain B content is explained supposing that, at this concentration, B begins to enter into the interstitial positions, or that a B_4C clustering begins.

Tombrel [22] found, for a similar deposition temperature, a similar dependence of the degree of graphitization on B content. The fact that his $\bar{g}$ values were higher than those found by Kotlensky could be explained by the longer deposition time used in Tombrel's experiments. Preferred orientation was found, however, to decrease as the B content was increased.

Medium temperature BPC deposited by Marinković et al. at 1670 to 1900 K [23] shows an improvement in the structural properties with increasing B content up to a maximum at about 1 percent B. The maxima were found for the degree of graphitization (Fig. 7), the preferred orientation of the crystallites, the crystallite height L_c, and the apparent density. In addition, the oxidation of BPC in a saturated solution of $Ag_2Cr_2O_7$ in concentrated sulfuric acid was studied following the method of Oberlin and Mering [24]. The first step of the oxidation consists in an insertion of sulfuric acid between carbon layers, thus making an intercalation compound; and therefore, the

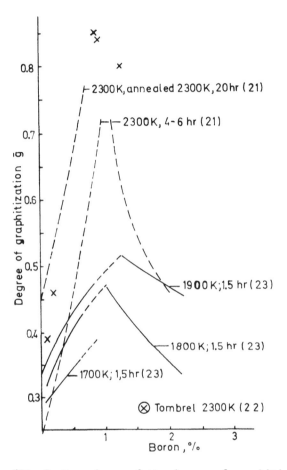

FIG. 7 Dependence of the degree of graphitization on B content, according to Kotlensky [21], Marinković et al. [23], and Tombrel [22]. Experimental points are omitted for clarity, except for Tombrel's results.

less defective the carbon structure, the higher its aptitude for the oxidation [24,25]. The undoped PC shows the presence of two components, one of which is oxidized at a rate about 10^3 times greater than the other. The volume fraction of the rapidly oxidized component is only about 20 percent.

Interestingly, although the oxidation rate of the rapidly oxidized component in the BPC increases with B content up to a maximum at about 1 percent B, its amount is less than in the undoped PC.

According to the structural model proposed by Mering and Maire [25,26], the rapidly oxidized component consists of pairs of carbon hexagon layers without interstitial atoms between them, while the slowly oxidized component contains the interstitial atoms. Consequently, Marinković et al. interpreted their own results assuming that in the process of deposition, the B atoms are distributed between substitutional and interstitial positions. At lower concentrations they preferentially occupy the substitutional sites, whereas at higher concentrations they go into the interstitial positions. This assumption could explain the maxima in the properties, but also the observed scattering of the data, this apparently being due to the uncontrollable variations in the distribution of B atoms between substitutional and interstitial sites.

Accordingly, the results of the oxidation measurements were explained supposing that the presence of B atoms in the substitutional sites increases the oxidation rate of the rapidly oxidized component, but the presence of interstitial B atoms impedes the reagent penetration, thus transforming the rapidly oxidized component into the slowly oxidized one.

To check these hypotheses, heat-treatment experiments were made with samples having different B concentrations [27]. The samples were heat-treated in vacuum at 1810 and 2090 K because, according to Turnbull et al. [7], these temperatures are sufficiently high to remove the interstitial B atoms, but they are too low to remove the substitutional B. The results show that HT leads to a decrease in B content of the BPC, which tends to a constant value in the range 0.6 to 0.9 percent B. This indicates the maximum B content retained in substitution.

Contrary to the behavior of PC on HT, the properties of which are only slightly changed, the properties of the BPC samples are considerably affected. The interlayer spacing becomes noticeably smaller and the amount of the rapidly oxidized component content increases somewhat. This could be expected because these properties should depend directly on the presence of interstitial B atoms. It is possible, however, that HT also leads to an increase of substitutional B atoms (see also Ref. 36), thus provoking the interlayer spacing decrease. An increase of substitutional B occurring together with a removal of interstitial B is also indicated by the fact that HT leads to a relatively small increase in the rapidly oxidized component content, which remains lower than that of the pure PC. Namely, an increased substitution would lead to a Fermi level lowering which would, in turn, lessen the sulfuric acid insertion and thus diminish the amount of the rapidly oxidized component.

At the same time, HT provokes a small modification of preferred orientation and crystallite height, because these depend most on the deposition conditions, and cannot be substantially changed by HT.

A striking difference in the behavior of mechanical properties between the BPC and the undoped PC, both deposited at 2320 K, was observed by Kotlensky and Martens [28] and Tombrel [22]. The BPC samples containing about 1 percent B were submitted by Kotlensky to tensile tests at high temperature and the structural properties of these samples were measured after the treatment. The fracture strain of the BPC was found to be similar to that of the undoped material up to 2470 K, but at higher temperatures the elongation of the BPC was much higher (Fig. 8). Such large deformations have induced structural transformations (e.g., degree of graphitization approached 1), which is not typical of the behavior of PC. Similar results were reported in Ref. 22.

The existence of minima and maxima on the curves representing structural properties of PC deposited at 2120 K versus its B content was also found by Katz and Gazzara [29]. The interlayer spacing c, microstrain η, and flexural strength σ, all show minima at or near 0.5 percent B, but at higher concentration c and σ increase up to about 1 percent and remain approximately constant above that concentration, while η, after passing through a minimum at 0.5 percent B, increases up to about 1 percent B, where a distinct maximum appears. Crystallite height L_c increases with B content up to a maximum at 0.75 percent B.

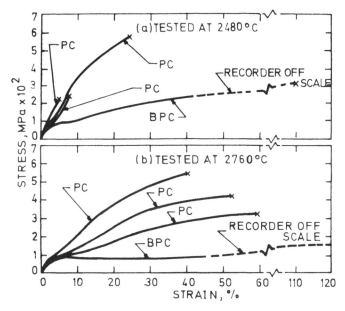

FIG. 8 Typical stress-strain curves for B-doped (BPC) and undoped
pyrolytic carbon (PC). (From Ref. 28.)

The authors essentially agree with Kotlensky's conclusion that
the existence of a maximum in the degree of graphitization is a consequence of the effect of substitutional and interstitial (or B_4C)
boron, which diminish and increase the interlayer spacing, respectively. However, as a limit of the substitutional solid solubility
of B, they state the concentration range between 0.4 and 1 percent B.
Boron carbide is supposed to form above the substitutional solubility
limit and its presence is confirmed in the sample containing 2.3 percent B. The presence of B_4C is supposed to cause severe microstrains
by forming a coherent interface with the carbon lattice for B contents
around 1 percent.

Klein was among the first researchers to make an extensive investigation of the transport properties (electrical resistivity, galvanomagnetic properties, Seebeck coefficient) of BPC. His results are
presented in Volume 2 of this series [30], and only a short summary
is given here. The BPC was deposited at a temperature of around
2270 K and its B content was in the range of 0.1 to 1 percent.

With increasing B content, electrical resistivity was found to decrease and to become almost independent of temperature, thus reflecting a drastic lowering of the Fermi level. At the same time, the Hall and Seebeck coefficients show that from a two-carrier type (with electrons predominating at high temperature) BPC is being converted to a pure p-type conductor.

Klein stated that the band theory provided a valid frame of reference for most of his experimental results and developed a quantitative interpretation of electron transport along basal planes, based on a simple two-band model assuming two parabolic energy bands with a very small, or zero overlap. A good agreement between experimental results and theoretical predictions was achieved, although the author anticipated that certain refinements would be necessary when dealing with some of the low-temperature processes.

Klein concluded that the Fermi level was already depressed by about 0.02 eV in "pure" PC due to crystal defects which trapped a certain number of electrons, while a B content of 1 percent led to a lowering of the Fermi level by about 0.1 eV.

A change of the Fermi level due to B substitution can be checked by studying the formation of lamellar compounds. The dual, amphoteric nature of a carbon hexagon layer in graphite or carbon, which can behave both as an electron donor and an electron acceptor, allows the formation of lamellar compounds with electron acceptors, such as halogens, or with electron donors, such as the alkali metals.

Of special interest is the insertion of Br and Na, because of their values of electron affinity (Br) and ionization potential (Na), which are critical regarding electron transfer to the carbon layers [26,31]. Thus the electron affinity of Br is just sufficient to allow extraction of electrons from the valence band of a carbon heat-treated above 1970 K [32]. Doping with B leads to a considerable lowering of the Fermi level and therefore, the higher the B content, the lower should be the absorption of Br. The results of Gremion et al. [32] show that absorption of Br drops sharply as the B content increases, so that BPC with about 1 percent B absorbs very little Br.

B doping also leads to an increased ability of carbon to absorb Na the inserted quantity of which increases with increasing B content in the carbon [31,33].

Structural and magnetic properties of BPC containing 0.2 percent B deposited at 2370 K were investigated extensively by Gasparoux et al. [34]. Three-dimensional ordering was found to take place upon HT at a considerably lower temperature than in the undoped PC, being virtually complete at about 2600 K, but the behavior of L_a was not very much different for the two carbons, indicating that B affects crystallite growth only slightly.

The diamagnetic anisotropy minimum occurring at a HTT of about 2970 K for the undoped PC apparently does not exist in the case of BPC (Fig. 9). The behavior of the anisotropy of diamagnetic suscep- tibility as a function of ambient temperature was found to differ from that of graphite, PC, or partially graphitized carbon (for HTT > 1870 K). Instead of a steady increase of the anisotropy with $1/T$, typical of these materials, the BPC shows a distinct maximum at about 1000 K, the position of which is shifted to lower temperature as HTT increases.

It is true that such behavior was theoretically predicted by McClure [12] starting from the SW model, and observed by Delhaes and Marchand on boronated carbons [35] and polycrystalline graphite [13]. However, the maximum for BPC is much sharper than in other carbons and is less readily explained by this theory. An upward shift of the curves with increasing HTT is, at the same time, readily explained by a lifting of the Fermi level, due to increased graphitization.

Difficulties in the interpretation of the diamagnetic anisotropy behavior arise partly from the inadequacy of the available models, but they may also be due to the possibility that the amount of sub- stituted B varies with HTT, which is indicated by the experimentally found heterogeneity in the samples.

An approximate comparison of the experimental results with McClure's theoretical curves gives a value of 0.1 eV for the lowering of the Fermi level in the as-deposited BPC, which is in agreement

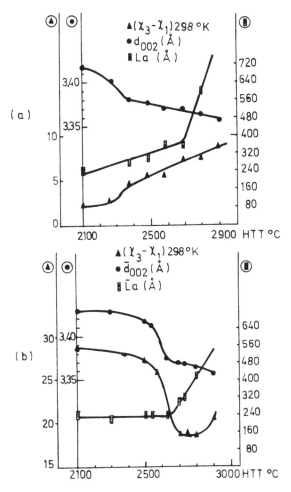

FIG. 9 Dependence of $(\chi_3-\chi_1)$, $\overline{d}_{002}$, and $\overline{L}_a$ on HTT for (a) B-doped and (b) undoped pyrocarbon. (From Ref. 34.)

with Klein's results [30]. It was experimentally shown, furthermore, that the diamagnetic susceptibility parallel to the layers, χ_1, and the mean disorientation angle of the crystallites, θ, do not depend on measurement temperature. The value of χ_1 was found to amount to 0.3 (i.e., to be close to the graphite value). At the same time the susceptibility values of the BPC were considerably lower than those for the pure PC, which is readily understood, taking into account the influence of B on the lowering of the Fermi level.

An investigation of the properties of BPC with B concentrations ranging from 0.04 to 0.16 percent [36] has shown that the Hall coefficient, electrical resistivity, and magnetoresistance are nearly independent of temperature, and decrease with B content. The characteristic maxima of the diamagnetic susceptibility versus 1/T curves have been found to become higher with increasing B content. This behavior of magnetic anisotropy was quantitatively interpreted using a simple two-dimensional electronic model.

This model had to be slightly modified, however, to account for the behavior of a similar series of pyrocarbons containing up to 0.10 percent B upon their heat treatment at 2570 to 2970 K for various times, up to 66 h [37]. The results indicate that the concentration of holes decreases rapidly at the beginning of the graphitization process and remains constant during the later stages. The authors conclude, therefore, that the efficiency of doping decreases as graphitization proceeds; that is, substitutional B atoms leave their positions in the lattice and migrate to form carbide clusters. This is in agreement with the already mentioned result [7] that substitutional B diffuses out of the lattice above 2300 K. The corresponding shifts of the Fermi level were explained assuming that the density of states decreased during the earlier stages of the graphitization process. The low value of the apparent activation energy of graphitization (150 to 170 kcal/mole) was attributed to the catalytic action of B.

The doping efficiency, found to be 30 to 60 percent [36], is much higher than that achieved by diffusion of B into a coke [35]. This is presumably due to the doping method, in which deposition of B takes place simultaneously with the formation of the carbon structure.

Maillard and Maire [38] have devised interesting experiments to clarify the influence of B on graphitization. By analyzing the modulation of the (112) x-ray reflection and by measuring the a and c parameters of a graphitizable carbon submitted to a graphitizing treatment with and without the presence of B and then progressively eliminating B by treatment with chlorine, they concluded that B catalyzed graphitization, but that the degree of graphitization should not be calculated from the interlayer spacing. Namely, B substitution causes the carbon layers to approach each other independent of the degree of three-dimensional ordering, in agreement with Refs. 7 and 8. Another conclusion is that partial elimination of B from the sample favors B substitution in the lattice (parameter a increases) and augments the degree of graphitization. However, a virtually complete (subsequent) elimination of B at a relatively low temperature (1870 K) leads to a "degraphitization." Only subsequent treatment at a higher temperature (2270 K) produces a "regraphitization" and a reestablishment of the validity of the $P_1 = \overline{g}^2$ relationship.

It seems that such a behavior could be related to an increased ability of the B to diffuse in the graphite lattice [17,39]. An increased degree of graphitization obtained by partial elimination of B at a relatively low temperature (1570 K) can be explained by an increased ability of B (with respect to carbon) to diffuse throughout the sample. Further treatment of the sample (at higher temperatures) leads to a degraphitization because of an almost complete elimination of B. This disturbs the lattice, the order of which cannot be restored, due to the removal of B, resulting in diminished diffusion. Only HT at a temperature sufficiently high to provoke self-diffusion of C atoms leads to the elimination of the structural defects, resulting in the regraphitization.

Delhaes and Marchand [35] have investigated the effect of B introduced into a partly graphitized coke on its electronic properties, with the aim of determining the efficiency of B substitution and the validity of the existing model of the electronic structure.

No difference in the degree of graphitization was found between the undoped and boronated (0.35 percent B) carbon heat-treated up to 2620 K, which induced the authors to conclude that B apparently did not affect the graphitization of the coke. As a matter of fact, the nominal B concentration should be high enough to noticeably increase the degree of graphitization, but the number of B atoms effectively substituted for C atoms is much lower (the efficiency of doping being only 1 to 10 percent), which explains the authors' observation.

The diamagnetic susceptibility (χ) versus measurement temperature curves of the boronated samples for any HTT lie below those of the undoped carbon. The Fermi level lowering due to B substitution leads to a lowering of χ. The greatest effect was found for a HT lasting 3 h at a given HTT (i.e., χ decreased up to 3 h and then increased again). The fraction of B atoms substituted for C atoms (i.e., efficiency of doping) was found to be highest for a HT time of 3 h. χ of the pure HT coke was found to behave in accordance with the two-dimensional free-carrier-gas model. However, the behavior of the B-doped carbon is in accordance with the model only for HTTs below 2250 K, while for higher HTTs some of the curves show maxima. Although such behavior is qualitatively predicted by McClure's calculations, they cannot be applied to strongly, but not fully graphitized carbons.

Interpretation of the diamagnetic properties by means of the model can still be regarded as satisfactory, but room temperature Hall coefficient values are in all cases too low to be interpreted by the model and there is a strong discrepancy in the Fermi level position as determined from diamagnetic susceptibility and from Hall effect values. Therefore, a modification of the model was considered on the basis of three carrier types, supposing that at any temperature there are a certain number of mobile electrons beyond the π band. In addition to the holes and π-band electrons, the existence of impurity band electrons is postulated to explain all the existing experimental data on magnetic and galvanomagnetic properties of carbons.

In connection with the doping mechanism, the authors point out that it is very difficult to achieve doping by B if graphitization does not occur simultaneously, and that optimum conditions for doping exist when the "amorphous" zones are being transformed into "graphitic" ones. The broken bonds caused by a removal of heteroatoms, or by broken tetrahedral bonds, may be occupied by B atoms. The authors' doping experiments, in which they impregnated the carbons with B and subsequently treated them at a HTT lower than the initial HTT, gave only a very small doping effect.

The numerous data concerning the effect of B on the properties of graphitizable carbons suggest the following conclusions:

1. Boron enters into the carbon lattice, improving the structural characteristics up to its solubility limit, found to be about 1 percent. Above this limit, B atoms seem to enter interstitial positions in the lattice, and have an inverse effect on structural properties.

2. There is evidence suggesting that during the formation of boron-containing pyrocarbon, the B atoms are distributed between substitutional and interstitial positions, preferentially occupying the substitutional or interstitial sites, depending on whether the B content is below or above the substitutional solubility limit.

3. There is no evidence suggesting that the mechanism of graphitization of B-containing carbon should substantially differ from that of the pure carbon.

4. The efficiency of B substitution is increased if B is introduced in the process of the formation of the carbon lattice, or during structural rearrangements.

5. B substitution leads to a depression of the Fermi level of the carbon, thus affecting its transport properties and the formation of intercalation compounds.

6. The diamagnetic susceptibility behavior of B-containing carbon, although qualitatively explained, cannot be accounted for quantitatively by the existing theoretical models.

C. Nongraphitizable Carbons

The carbons of this class, often referred to as hard carbons, consist of small elementary domains completely disoriented with respect to each other and strongly bonded by covalent C-C bonds. Only a small part of such carbons (10 percent) can be transformed into graphite, while most of their mass does not undergo any transformation upon HT, because the thermal energy is not sufficient to break the strong carbon-carbon cross-links. An addition of higher metal concentration

(5 to 10 percent) has a favorable effect on graphitization ("catalytic graphitization").

Rouchy and Mering [40] have studied graphitization of hard carbons obtained from saccharose. They used B additions too low (below 1 percent B) to permit a "catalytic graphitization" (dissolution-precipitation) mechanism to become operative.

As a method of doping, pyrolysis of triphenylborane was used, either during carbonization, performed by heating the saccharose at 1770, 1270 or 620 K, or after the carbonization by heating the pre-treated saccharose in the presence of the organoborane. The organoborane was used as a source of B in order to avoid the introduction and possible influence of other elements, such as Cl or O.

For the given doping conditions, increasing B content (of up to 0.8 percent) led to a progressive transformation of the pretreated coke to graphite. However, a B concentration above 1 percent leads to the addition of the dissolution-precipitation effect (B_4C suddenly appears above 1 percent B) to the effect of doping.

If B is present during carbonization, a much more complete transformation of the coke to graphite can be achieved. However, even the x-ray pattern of the coke treated to the highest HTT (2770 K) contained the pattern of the residual coke superimposed on fine reflections of graphite.

The authors' explanation of the observed effects of B is as follows. The introduction of B into the carbon structure creates electronic defects and considerably changes the electronic properties of the hard carbon. The bonding between the small domains of the coke becomes less rigid upon heating, their mobility is increased and their mutual rearrangement facilitated. The hard carbons can be thus graphitized by HT at 2770 K.

However, very dispersed B atoms can change only the local structure of the coke, and the transformation is possible only by B diffusion at high temperatures. The graphite thereby formed reacts with the B atoms which become incorporated in the graphite lattice. In this way the number of B atoms available for further diffusion and

reaction with carbon diminishes, finally becoming zero. The transformation is therefore never complete.

III. SILICON

Silicon is the nearest neighbor to carbon in the fourth column of the
periodic chart of the elements. Both elements have an analogous outer
electron structure with two s and two p electrons, which determines
their similar chemical behavior. In both elements sp^3 hybridization
is preferred, giving four equal, strictly directed chemical bonds,
so that there are analogous compounds of carbon and silicon. This
electron configuration is also preserved in the structure of the
solid elements, resulting in the so-called diamond structure, which
is also found in the stoichiometric compound between Si and C, silicon carbide.

However, the chemistry of carbon and the chemistry of silicon
are only partly alike. The differences in chemical behavior of these
two elements are mainly caused by the special ability of carbon, not
characteristic of silicon, to form multiple bonds. The sp^2 hybridization which produces three equivalent coplanar σ bonds does not
involve the fourth electron. This electron is delocalized and forms
the relatively weak π bonds. Such bonds are found in aromatic compounds which have a planar molecular structure and in graphite, which
can be regarded as the extreme member of the series of aromatic compounds of increasing molecular weight [41]. In silicon such π bonds
and, consequently, multiple bonds are not stable, and therefore the
silicon compounds with compositions corresponding to those of unsatu-
rated carbon compounds are stable only as polymeric structures (i.e.,
without double bonds).

As a result, pure solid silicon with not all of the bonds mutually equivalent, and having a structure analogous to that of graphite does not exist. Only in certain solid compounds, such as inter-
metallic compounds between silicon and some transition elements, are
silicon substructures in the form of hexagonal layers known.

Bearing in mind the properties of silicon, the question of the possibility of its substitution for carbon atoms becomes very interesting.

A. Structural Properties

There was virtually no work on the solid solubility of Si in carbon until the late sixties, when Emyashev and Lavrova [42] and Lavrov et al. [43] reported that by copyrolysis of a CH_4-$SiCl_4$-H_2 mixture at 2100°C, "alloys" of carbon with silicon could be prepared. The structural properties (interlayer spacing c/2, crystallite sizes L_a and L_c) of the alloyed PC were improved up to 0.1 weight percent Si, but preferred orientation was less in the Si-alloyed material.

In the work of Yajima and Hirai [44-47] on the chemical vapor codeposition of carbon and silicon from a propane-$SiCl_4$ mixture at 1710 to 2300 K, an improvement of structural properties of the PC which contained up to 4 weight percent Si was observed. The Si was predominantly present as preferentially oriented β-SiC particles.

In the extensive early work on chemical vapor codeposition of carbon and silicon in the author's laboratory [48,49], a PC containing Si was prepared in the temperature range 1370 to 1900 K from a CH_4-$SiCl_4$ gas mixture. Most of the experiments were performed at 13.3 kPa total pressure using 2000 ml/h CH_4 flow rate.

The maximum Si content that can be retained in the deposited carbon was found to depend strongly on the deposition temperature (Fig. 10). A sharp maximum was found to exist near 1570 K. The lowest temperature at which $SiCl_4$ decomposition could be observed was 1430 K, which can be compared with the results of thermodynamic calculations showing that the reaction of $SiCl_4$ with CH_4 to form SiC start at about 1200 K. Thus the low Si concentration at the lowest temperature is probably associated with the low rate of decomposition of $SiCl_4$.

A study of Si distribution by means of an electron microprobe analyzer showed that it was homogeneous only in the samples with low Si content (up to ∿0.3 percent Si), such as those formed at 1900 K, and in the samples with a higher Si content formed at the lowest

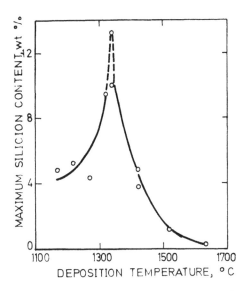

FIG. 10 Maximum silicon concentration in the deposit versus deposition temperature. (From Ref. 48.)

deposition temperature (below 1590 K). In the samples with a high
Si content (above 1.5 percent Si) formed at temperatures higher than
1590 K and having an uneven Si distribution, there is a correlation
between Si content and microstructure of the deposit. In addition
to regions of low Si content (up to 0.3 percent Si) characterized by
narrow growth cones usually somewhat fibrous in appearance, there
are large cones consisting of a SiC-PC mixture which apparently grow
from pure, lens-shaped SiC inclusions (Fig. 11). Within these large
conical formations there is a concentration gradient of Si: the maximum Si concentration is at the lens-shaped origin of the cones, where
it amounts to about 70 percent (i.e., corresponds to the SiC composition [50].

The SiC inclusions serving as nuclei for the large growth cones
seem to be carburized droplets of pure Si originally formed by the
reduction of $SiCl_4$. The droplets are not formed at the lowest temperatures (i.e., below the melting point of Si), which explains the
absence of such large cones in the deposits formed at these temperatures.

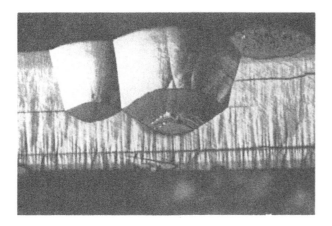

FIG. 11 Microstructure of the silicon-rich deposit (10 percent Si, ×150, polarized light). (From Ref. 48.)

The dependence of certain structural properties of the Si-containing PC deposited at 1800 K on its Si content shows distinct maxima at about 0.2 percent Si (Fig. 12). Such maxima are found for apparent density, preferred orientation, and crystallite height L_c, but also for the amount of the component which is rapidly oxidized in a saturated solution of $Ag_2Cr_2O_7$ in H_2SO_4. According to Mering and Maire [25,26], this can be taken as a measure of the amount of the graphitic layer content in the pyrocarbon product.

In addition to these results, an examination by electron diffraction revealed a distinct effect of Si on the graphitization of the PC. The two-dimensional (11) reflection appearing in the x-ray diffraction pattern of pure PC is split for samples containing 0.1 to 0.2 percent Si into the (110) and (112) reflections. Electron diffraction also shows that larger crystallites appear at this Si concentration. On the other hand, neither x-ray nor electron diffraction showed any other phase except PC. Although these results clearly point to the conclusion that the presence of Si improves structural features of the PC, interlayer spacing measurements did not show any obvious variation with Si content. This is an important difference in the behavior of the Si-containing PC from the boron-containing material and could not be explained by the larger atomic radius of Si, because

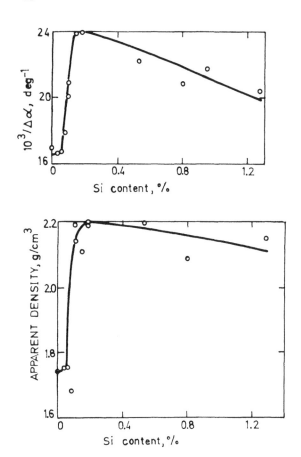

FIG. 12 Preferred orientation and apparent density versus Si content
(Δα = half-width of the distribution curve). (From Ref. 48.)

the Si concentration is too low. Interestingly, the x-ray (00ℓ)
reflections of the Si-containing PC showed a distinct asymmetry
(Fig. 13). This fact was the starting point of a more detailed
investigation of x-ray (00ℓ) reflections with a new series of Si-
containing specimens codeposited at 1900 K [51]. Since in some
cases the (004) profiles showed a distinct asymmetry, indicating
a superposition of two reflections, a detailed study of the effect
has been undertaken. It has been found that all the Si-containing
PC samples give asymmetrical (004) profiles with respect to the

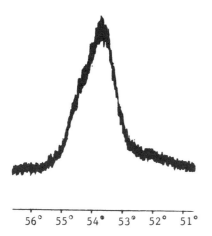

56° 55° 54° 53° 52° 51°

FIG. 13 X-ray diffraction (004) profile of a pyrocarbon containing Si (0.17 percent Si, HTT 2300 K).

normal to the background passing through the maximum of the reflection, whereas there is no asymmetry in the case of the undoped PC. The degree of the asymmetry was found to depend on the Si concentration in a manner similar to that found for other structural properties; that is, it increased with Si content up to a maximum appearing at a given Si concentration (0.14 percent) and decreased above this concentration. The (00ℓ) reflections were found to be deformed toward the larger angle, which would indicate the presence of a more graphitized component besides the turbostratic one. Since the asymmetry provoked errors in the half-width of the reflections from which the crystallite size L_c was determined, the half-width was determined from the low-angle part of the divided (002) reflection.

The plot of the L_c values thus obtained versus Si content was again similar to those of the other properties, showing a maximum at 0.14 percent Si (Fig. 14). It was then assumed that the asymmetrical SiPC reflections were composed of two closely spaced symmetrical profiles, each corresponding to a definite structural component. The next step was to separate the (004) profiles into their assumed components in order to determine their relative amounts. This separation is facilitated by the fact that the (00ℓ) profiles

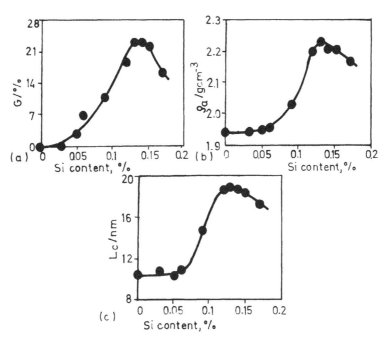

FIG. 14 Properties of SiPC versus Si concentration: (a) graphitic component content G; (b) apparent density; (c) crystallite height. (From Ref. 51.)

from the undoped PC are symmetrical, except for a small deformation at low angles. Therefore, an additional assumption was introduced postulating that both component profiles were symmetrical and that the low-angle part of the profile [i.e., the one corresponding to the predominant (turbostratic) component] was unaffected by the presence of the other component.

In addition to the graphical separation, another approach was used, which consisted of approximating the predominant profile by an expression representing an intermediate case between a Lorentzian and a Gaussian function [52]: namely

$$ I_0 = \frac{I_m}{1 + ax^2 + bx^4} \qquad x = \frac{\theta - \theta_m}{B/2} \ . $$

I_0 is the intensity of the predominant component at the angle θ, I_m is its maximum intensity, a and b are parameters determined by fit-

ting the calculated profile to the experimental one in the region
where the influence of the other component can be neglected, θ_m is
the profile-maximum angle, and B is the profile half-width.

The separation, which in both procedures gave similar results,
showed that in addition to a turbostratic component (having an interlayer spacing of 0.3414 nm and thus similar to the undoped PC), there
was another, graphitic component characterized by an interlayer spacing of about 0.336 nm. Since both component profiles have similar
widths, their relative amounts were calculated from the corresponding
peak heights. Fig. 14-a shows the dependence of the graphitic component content of the SiPC on Si concentration. It increases with
Si concentration up to a maximum value of 24 percent, situated at
0.14 percent Si, and decreases above that Si concentration. Thus,
again a similar curve is obtained as for the other properties.

At this point a comparison can be made of the Si-containing PC
with the B-containing material:

> The properties of the material in both cases show similar qual-
> itative dependence on the element concentration, with maxima
> situated at about 1 percent B and 0.14 percent Si, respec-
> tively.

> The positions of the maxima differ considerably, particularly
> on the atomic scale: in the case of Si they appear at about
> 0.06 atomic percent instead of about 1 atomic percent B.

Although the presence of both elements improves the structural
characteristics of the resulting carbon, there is a qualitative difference in the influence of the two elements. Instead of a single-
component turbostratic structure typical of undoped and B-containing
PC, doping by Si results in a two-component (turbostratic and graphitic) structure. In other words, the difference in the influence of
Si and B is in the number of successively stacked graphitic layers.
In the B-doped carbons (as well as in the undoped partly graphitized
carbons) pairs of graphitic layers are randomly scattered throughout
the material [26], so that their increasing number (increased graphitization) is manifested only in the angular positions of the (00ℓ)

reflections, which are shifted toward a higher angle. However, in the Si-doped carbon there is a distinct graphitic component consisting of some 40 to 50 graphitic layers stacked together, and increased graphitization consists of an increase of the number of such "crystallites." This is manifested by a greater deformation of the high angle end of the (00ℓ) profiles and results in a higher graphitic component peak (after the peaks corresponding to the two components are resolved).

B. Graphitization Behavior of Silicon-Doped Pyrocarbon

In order to follow the graphitization behavior of the Si-doped PC, it was subjected to HTT together with an undoped material deposited under similar conditions [51]. The structural characteristics were measured by the methods already described.

Table 1 shows interlayer spacing c/2 and crystallite height L_c of both Si-doped and undoped PC for various HTTs and HTts (heat-treatment times). The data for the undoped PC show a gradual shift of the (00ℓ) peaks, although the effects are small. The crystallite size remains roughly constant upon HT, the changes, if any, being very small.

TABLE 1 Interlayer Spacing and Crystallite Height in PC and SiPC Subjected to Various Heat-Treatment Conditions

T_{ht} (K)	t_{ht} (min)	PC		SiPC			
				Turbostratic component		Graphitic component	
		$c/2$ (004) (nm)	$L_{c,(002)}$ (nm)	$c/2$ (004) (nm)	$L_{c,(002)}$ (nm)	$c/2$ (004) (nm)	$L_{c,(002)}$ (nm)
As deposited		0.3414	12.4	0.3410	13.4	0.3363	
2300	30	0.3411	12.9	0.3405	14.2	0.3360	
	60	0.3408	14.6	0.3400		0.3359	
	90	0.3408	12.4	0.3400		0.3360	16.5
2400	30	0.3408	12.4	0.3400	13.1	0.3364	
	60	0.3405	14.1	0.3399		0.3366	17.0
	90	0.3402	14.1	0.3399		0.3362	17.4
2600	30	0.3399	13.0	0.3409	14.6	0.3369	
	60	0.3390	13.2	0.3401		0.3358	17.2
	90	0.3389	13.0	0.3402		0.3360	18.0

Source: Ref. 51.

The SiPC samples show composite (00ℓ) reflections. The structural changes caused by HT were followed by separating the (004) profiles into the two components mentioned. Since the profiles of the components overlap, the separation procedure is susceptible to significant errors, particularly if the peaks have similar magnitudes.

The results presented in Table 1 show that the interlayer spacing of both components is not changed significantly upon HT, and the same is true for the crystallite height L_c.

The principal effect of HT observed in SiPC is a change in the relative amounts of the two components, as deduced from the areas of the respective (004) profiles (Fig. 15). This figure shows the dependence of the graphitic component content, G, on the initial Si concentration for the as-deposited and heat-treated samples. The greatest effect is found in the sample with 0.13 percent Si (i.e., having the concentration at which maxima in the properties occur). Lower (or higher) Si concentrations produce smaller effects. No effect [i.e., no asymmetry of the (00ℓ) profiles] is found in undoped PC.

It is interesting to compare the effect of HT on graphitization of Si-containing and pure PC. The degree of graphitization $\bar{g}$, defined, according to Maire and Mering [26], by the equation

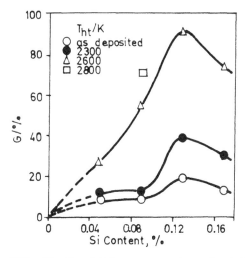

FIG. 15 Graphitic component content G versus initial silicon concentration for as-deposited and isochronally heat-treated SiPC. (From Ref. 51.)

$$\overline{g} = \frac{0.344 - (c/2)_{002}}{0.0086}$$

could be determined in the Si-doped PC, taking the observed fractions of the graphitic and turbostratic components and their respective interlayer spacing values (Table 1). Thus, for the sample with 0.13 percent Si, HTT = 2600 K, we have

$$\overline{g}_{0.13\%Si}^{2600\ K} = \frac{0.344 - (c/2)_G}{0.0086}\ G + \frac{0.344 - (c/2)_T}{0.0086}\ (1 - G) = 0.90$$

where $(c/2)_G = 0.336$ nm and $(c/2)_T = 0.340$ nm (Table 1); G = fraction of the graphitic component = 0.94 (Fig. 15). This is to be compared with the degree of graphitization of the undoped pyrocarbon HT under identical conditions, $\overline{g} = 0.58$.

An additional point that deserves attention is the value of c/2 of the graphitic component, which appears to be about 0.336 nm. (Although the separation of the peaks introduces errors in the peak positions, the values in Table 1 concentrate around 0.336 nm.) According to Fischbach, who studied graphitization of undoped PC [3], this value corresponds to a relatively stable imperfect structure reached after sufficient HTt at temperatures up to 3300 K. The graphite value of c/2 = 0.3354 nm was not observed in Fischbach's isothermal HT studies, even after HTts of the order of 10^3 min at 3300 K, but could be reached at temperatures above 3300 K. On the other hand, those studies show that to reach a c/2 value of 0.336 nm within the time interval used in our studies (up to 90 min), a temperature of about 3200 K would be necessary.

In addition to the analysis of the (00ℓ) profiles, the modulation of the (hk) reflections was also investigated. The effect of heat treatment on the (hk) reflections of both SiPC and pure PC is shown in Fig. 16. A striking difference in modulation is observed. The modulation of the SiPC is noticeable even in the as-deposited material, and it is quite pronounced after HT, whereas the undoped PC hardly shows any modulation even after the extreme HT conditions used in this study.

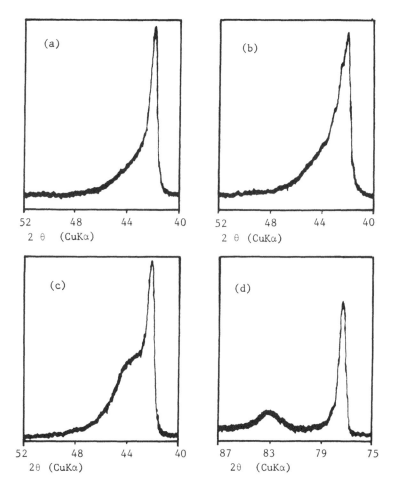

FIG. 16 The (hk) reflections (CuKα radiation) of PC and SiPC after various heat treatments: (a) PC treated at 2600 K for 75 min, (10) reflection; (b) SiPC untreated, (10) reflection; (c) SiPC heat-treated at 2600 K for 75 min, (10) reflection); (d) SiPC heat-treated at 2600 K for 75 min, (11) reflection. (From Ref. 51.)

The apparent modulation of the (hk) reflections in the SiPC may be a consequence of two effects: (1) a contribution from the graphitic component, assuming the presence of such a component on the basis of the analysis of the (00ℓ) reflections; and (2) a contribution from ordering in the turbostratic component.

Although the interlayer spacing of the turbostratic component appears to decrease slightly upon HT, the effect is small and could produce only a negligible modulation of the (hk) profile. Thus it follows that the apparent modulation is in fact a superposition of the (hkℓ) reflections from the graphitic component on the (hk) pro- file from the turbostratic component.

On the basis of these results, it can be concluded that the graphitization of SiPC occurs by a mechanism different from the graphitization mechanism of the pure PC. Instead of the progressive ordering of the turbostratic structure typical of pure PC, a direct conversion of the turbostratic component into the graphitic one occurs in the SiPC.

A comparison of the behavior of the PC and SiPC upon HT shows that the specific graphitization process of the latter is also faster. In order to obtain more data about the kinetics of the process, ex- periments were made to establish the dependence of the turbostratic component content on HT time. Figure 17 shows a plot of the loga- rithm of the ratio of the amount of the turbostratic component after

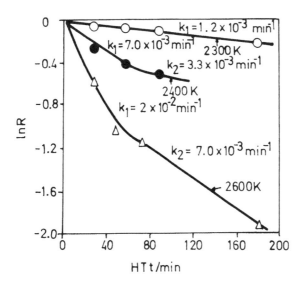

FIG. 17 Logarithm of the ratio R—the amount of turbostratic com- ponent after a heat-treatment time (HTt) relative to the initial amount—versus HTt, for different HTTs. (From Ref. 51.)

a heat-treatment time (HTt) to the initial amount of that component
as a function of HTt for temperatures 2300, 2400 and 2600 K. The
relationship is represented by a single straight line for 2300 K,
but for higher temperatures two straight lines with the rate constants differing by a factor of 2 to 3 were obtained.

It is interesting to compare the graphs in Fig. 17 with Fischbach's results relative to the graphitization of PC [3]. Fischbach
plotted the logarithm of the fractional change in the interlayer
spacing of the single-component undoped PC deposited at 2420 K versus
HTt and obtained graphs qualitatively similar to those presented in
Fig. 17. One of the most interesting conclusions from his results is
that "the graphitization develops in a succession of distinct steps
or stages": stage I, when the interlayer spacing decreases from 0.343
nm to 0.337 nm and where layer ordering is the dominant process, and
stage II, where the interlayer spacing decreases further, to about
0.336 nm, and in which the increasing crystallite size L_a plays a
very important role.

It would be logical to suppose from the data presented in Fig.
17 that graphitization of the SiPC consists, similar to the case of
high-temperature PC, of two mutually time-shifted processes with
considerably differing rate constants. It seems clear, however,
that the graphitic component grows at the expense of the turbostratic
one (i.e., that the latter is transformed into the former).

A comparison of the rate constants for a given HTT shows that
the SiPC is graphitized much faster than the high-temperature PC.
Another interesting point is the comparison of the activation energies of the two processes. Again, the activation energy for SiPC
graphitization, although not known with sufficient accuracy, is considerably lower than the values obtained for the high-temperature PC
[3,53-56].

Thus the presence of a small concentration of Si in the PC
causes a graphitization process which apparently consists in a conversion of the existing turbostratic "crystallites" into the graphitic ones. This conversion is progressive in the sense that the
material becomes gradually richer in the graphitic component at the
expense of the turbostratic one. The process is also statistical.

However, within a given crystallite of the turbostratic component the conversion is not progressive but sudden; that is, the
structural rearrangement occurs suddenly in the whole body of the
crystallite. That all the layers in the turbostratic crystallite
are suddenly rearranged so as to adopt the mutual positions characteristic of graphite is shown by the very fact that two distinct
components with definite values of interlayer spacing and crystallite
size (number of the layers stacked) are found in all the Si-containing
samples, irrespective of their Si content and the HT conditions to
which they were subjected.

The question of the mechanism by which Si can produce such a
difference in graphitization behavior is unsolved. It seems very
unlikely that the mechanism would be similar to that proposed for
the action of Ti (dissolution-precipitation), which has to be present
in a much higher concentration (5 percent) to produce a similar effect
[57]. The process that would be expected in the case of Si should
rather be connected with its presence in the substitutional sites of
the lattice and possibility to its diffusion in the structure.

C. Electronic Properties

Further insight into the structure and, in particular, the positions
and properties of Si atoms in PC may be obtained by investigating its
electronic properties. Only a few papers have been published on the
magnetic and galvanomagnetic properties of SiPC [58-60]. The SiPC
samples used in these studies were obtained under conditions used
for the deposition of the samples for structural investigations [51]:
a deposition temperature of 1900 K and a 1:1 CH_4/Ar gas mixture under
a total pressure of 13.3 kPa. These conditions ensured a maximum Si
concentration in the SiPC below 0.2 percent to avoid the formation of
SiC.

The galvanomagnetic properties were measured by a classical d.c.
method with parallel piped samples (15 x 3 x 0.3 mm) and electrical
contacts made by the electrolytic deposition of copper. The experimental setup enabled the authors to make measurements with a variable
magnetic field of up to 1.1 T [61].

Direct measurements of the diamagnetic anisotropy $\Delta\chi = \chi_\perp - \chi_\parallel$ ($\chi_\parallel$ and $\chi_\perp$ are susceptibilities measured parallel and perpendicular to the deposition surface, respectively) were performed using the critical angle method. In addition, the $\chi_\parallel$ and $\chi_\perp$ values were measured independently by the Faraday method. Assuming χ_1 to be independent of temperature, it was possible to obtain from the temperature variation of $\chi_\parallel$ and $\chi_\perp$ the values of the principal components of the susceptibility tensor of the carbon crystallites, χ_1 and χ_3. This calculation was made using three different methods [62-64], all of which gave results which agreed within the limits of a few percent.

The temperature range used for these studies was 4.2 to 300 K for the galvanomagnetic properties and 77 to 300 K for the diamagnetic properties. In some cases the latter measurements were made at a higher temperature (up to 650 K).

1. Galvanomagnetic Properties

The results are presented in Fig. 18, where the Hall constant, R_H, is given as a function of Si content and measurement temperature.

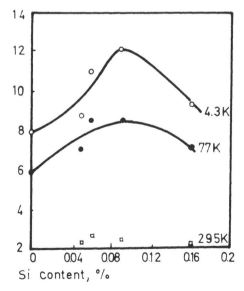

FIG. 18 Dependence of Hall coefficient on Si content and ambient temperature. (From Ref. 58.)

The positive values of the Hall coefficient indicate that the majority charge carriers are holes, rather than electrons, or at least
that electrons are less mobile than holes. The result that the Hall
constant decreases with measurement temperature is consistent with a
supposition that the majority of holes are charge carriers and that
there is a thermal excitation of electrons to the conduction band.

From Fig. 18 it also follows that the Hall coefficient increases
with Si content up to a maximum of about 0.1 percent Si. Measurements
of resistivity show that it is almost insensitive to the presence of
Si (apparently it slightly decreases with increasing Si content), and
that there is a slight decrease with measurement temperature. From
the Hall coefficient and electrical resistivity data, the Hall mobility of the charge carriers is calculated. Since resistivity is hardly
changed with Si content, the behavior of the mobility is mainly determined by that of the Hall coefficient. The increase of mobility with
Si content is quite evident, again with a maximum at about 0.1 percent Si.

The magnetoresistance values are generally small, of the order
of 0.1 to 1 percent. Figure 19 presents the dependence of the transverse component of magnetoresistance on Si content and measurement
temperature. At room temperature, pure PC shows a negative magnetoresistance, the magnitude of which increases with magnetic field
strengths in the range studied (i.e., up to 1.1 T). The sample with
0.05 percent Si has very small negative values for the field strength
below 0.6 T, and above this value the magnetoresistance becomes positive. For samples with higher Si contents, the magnetoresistance is
always positive and increases with the field strength.

At lower temperatures (liquid nitrogen and liquid helium temperatures) the magnetoresistance is always negative. The lower the
measurement temperature, the more negative the magnetoresistance.
The behavior of the longitudinal component of magnetoresistance measured at low temperature, which is several times lower and always
negative, is similar to that of the transversal component.

Although it is generally assumed that the dependence of the
magnetoresistance (ρ) on the magnetic induction (B) can be repre-

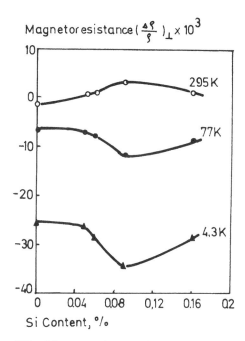

FIG. 19 Dependence of transverse component of magnetoresistance on Si content and ambient temperature. (From Ref. 58.)

sented by the function $(\Delta\rho/\rho) \sim B^n$ with $n \sim 2$, the values of n found in this work were 1.85 for room temperature and liquid N_2 temperature, and 1.15 for liquid He temperature.

2. *Diamagnetic Properties*

The anisotropy and transverse component of the diamagnetic susceptibility show a similar behavior with Si content as was found for other properties, with a maximum at about 0.1 percent Si (Fig. 20). The absolute value of the anisotropy increases with decreasing measurement temperature, so that the ratio of the values at liquid N_2 and at room temperature is about 1.4.

Comparison of the values of the diamagnetic anisotropy directly measured with those calculated from the measured $\chi_\perp$ and $\chi_\parallel$ components (Fig. 20) shows satisfactory agreement.

The values of χ_1 and χ_3, the longitudinal and transverse components of the diamagnetic susceptibility tensor of the crystallites,

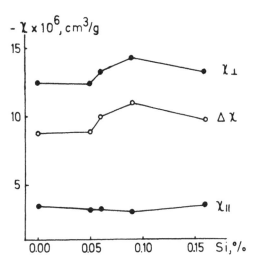

FIG. 20 Dependence of anisotropy and longitudinal and transverse components of diamagnetic susceptibility on Si content (at room temperature). (From Ref. 58.)

have been calculated from the temperature variation of $\chi_\perp$ and $\chi_\parallel$. The same methods of calculation yield the mean value of the angle θ between the c axis of the crystallites and the perpendicular to the deposition surface. θ is found to be close to 30°, similar to the values determined by x-ray diffraction. Although the calculated values differ somewhat from the experimental ones, the dependence on Si concentration is similar, showing that preferred orientation is a maximum at about 0.1 percent Si.

An interesting observation is that all of the values of χ_1 are considerably higher than those corresponding to the high-temperature PC, and virtually independent of Si content. Another interesting feature is that the values of the transverse susceptibility of the crystallites, χ_3, show a monotonous change with Si content. This is different from the behavior of $\Delta\chi$ and $\chi_\perp$ where a maximum appears at about 0.1 percent Si.

In order to follow the evolution of the diamagnetic properties, two of the samples were heat-treated for 3 h at 2300, 2600 and 2800 K. The samples chosen for the study were the pure PC and the 0.09

percent SiPC, the latter because the maxima in the majority of the properties investigated occur at that Si concentration.

The more important results are presented in Fig. 21. The anisotropy $\Delta\chi$ of the pure PC increases steadily throughout the entire temperature range, while the 0.09 percent SiPC reaches a maximum value of anisotropy at 2300 K and slightly decreases at higher temperature. The components $\chi_\perp$ and $\chi_\parallel$ show maxima at 2600 K (Fig. 21). It can be noted, however, that all the values tend to converge at high temperatures; that is, the values for PC and for the SiPC approach each other with increasing HTT. The same is valid for the χ_1 and χ_3 components, calculated from the temperature variation of $\chi_\parallel$ and $\chi_\perp$.

The Fermi level position and its dependence on Si concentration and HTT have been studied [60]. A model of a two-dimensional free electron gas, proposed by Marchand [65] and based on Stoner's treatment of the temperature dependence of free electron susceptibility [66], enabled the authors to calculate the energy difference ε_0 between the top of the valence band and the position of the Fermi

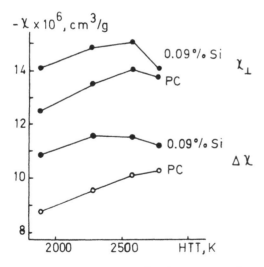

FIG. 21 Dependence of anisotropy and transverse component of diamagnetic susceptibility on HTT (at room temperature). From Ref. 58.)

level. According to the model, the temperature dependence of the
Landau magnetism is given by the expression

$$K = K_0(1 - e^{-(\varepsilon_0/kT)})$$

where K is the Landau magnetism (the anisotropic susceptibility of
free electrons), K_0 is the Landau magnetism at 0 K, and k is Boltz-
man's constant. Since it is more convenient to work with average
susceptibility $\overline{\chi}$, which is readily measurable and is related to K,
the expression above can be rewritten in the form

$$\overline{\chi} = C_1 - \frac{K_0}{3} e^{-(\varepsilon_0/kT)}$$

where C_1 is a constant composed of susceptibility terms and is independent of temperature. Consequently, if $\ln(1 - \overline{\chi}/C_1)$ is plotted
against 1/T, a straight line having a slope equal to ε_0/K should be
obtained. By fitting functional parameters to achieve an agreement
between calculated and experimentally obtained values of $\overline{\chi}$, the ε_0
values can be estimated.

The results obtained in this way are given in Tables 2 and 3.
Table 2 contains values of ε_0 versus Si content for the as-deposited
PC. The values decrease with increasing Si content, indicating the
lifting of the Fermi level. There is no indication of a minimum in
the vicinity of 0.1 percent Si.

Table 3 presents the ε_0 values for the two HT samples as a
function of HTT. The values of ε_0 decrease (i.e., the Fermi level
increases) with HTT for both PC and SiPC. The ε_0 values are lower
for SiPC, although the difference becomes smaller at the highest HTT.

The results pertaining to the galvanomagnetic and magnetic properties seem to be in favor of Si acting as a graphitization promotor.

TABLE 2 Dependence of ε_0 on Si Content

Si (%)	0.00	0.05	0.06	0.09	0.16
$\varepsilon_0 \times 10^2$ (eV)	3.00	2.73	2.68	2.61	2.54

Source: Ref. 58.

TABLE 3 Dependence of ε_0 on HTT for PC
and SiPC Containing 0.09 Percent Si

HTT (K)	Si (%)	$\varepsilon_0 \times 10^2$ (eV)
As deposited	0.00	3.00
	0.09	2.61
2273	0.00	2.72
	0.09	2.39
2573	0.00	2.65
	0.09	2.20
2773	0.00	2.29
	0.09	2.08

Source: Ref. 58.

The behavior of the Hall coefficient indicates a majority of holes
and thermal excitation of electrons into the conduction band with
increasing temperature. Assuming that this excitation is negligible
at 4 K, the Hall constant values would yield concentrations of charge
carriers of the order of 8×10^{18} cm^{-3} for the undoped PC, and 5×10^{18}
cm^{-3} for the 0.09 percent SiPC. This is quite a reasonable magnitude
for this kind of material, and the variation with Si content shows
that the presence of Si diminishes the concentration of electron
acceptors or traps with an approximate efficiency of 2 to 3 percent
(a return of two or three electrons to the valence band per 100 Si
atoms in the carbon).

The raising of the Fermi level, which is also indicated by the
behavior of the diamagnetic susceptibility, is quite consistent with
an enhancement of graphitization, as indicated by structural properties. The increase in the Fermi level might be also taken as an
indication that Si acts as an electron donor, but this seems highly
unlikely. In this connection it may be interesting to point out that
the Fermi level shift caused by B is much greater than that due to Si.

Silicon can thus act indirectly (as a graphitization promotor)
on the electronic band structure, and the observed increased mobility
and diamagnetic anisotropy are explained as normal consequences of an
upward movement of the Fermi level and increased graphitization of

the PC. The effect of Si on magnetoresistance also seems to be due
to increased graphitization.

The question might arise, however, as to why the maxima in the
electronic properties occur at 0.09 percent Si and not at 0.14 percent Si, as is the case with the structural properties, or why there
are no maxima or minima as far as some properties are concerned.
This question remains open to further investigation, although the
explanation may be very simple. Namely, the samples available for
measurement of the electronic properties did not cover the investigated concentration range with a sufficient number of points, so
that there is no "point" between 0.09 and 0.16 percent Si, that is,
in the very interval where the maxima in structural properties were
found to occur. Also, the concentration 0.16 percent Si is rather
close to that where maxima (minima) appear and the property values
at that concentration may or may not be influenced by such a neighborhood.

A question that is certainly not so simple is concerned with
the relatively high $|\chi_1|$ values exhibited by all the samples investi-
gated. These values are at least twice as large as would be expected
for any carbon material. Also, there is no obvious correlation with
the presence of Si. The answer may lie in the angular distribution
of the crystallites, which is, upon calculation of χ_1, assumed to be
independent of temperature. In fact, a very small variation of the
mean angle θ may be sufficient to alter significantly the value of χ_1.
A possible explanation may then be that χ_1 might have the same value
as in other carbons, but the thermal expansion from 77 K to 300 K
might be slightly different, leading to a temperature dependence of
the angular distribution of crystallites.

The effect of HT is, at least qualitatively, again consistent
with the conclusion that Si acts as a promotor of graphitization.
The quantitative explanations are not as simple. Questions arise as
to why there are maxima in properties for a certain HTT and why the
values of the properties for the pure PC and SiPC converge with HTT.
Part of the answer may lie in the fact that not all of the Si content
present initially is retained in the samples during HT, although the

lack of complete experimental data on the remaining Si content in
the heat-treated SiPC prevents a full discussion of the subject.

Although a general picture of the effects of a low Si content
on the properties of medium temperature PC is thus taking shape,
some important points remain obscure. The mechanism by which complete turbostratic crystallites are transformed into the graphitized
state remains to be explained.

In this connection, it would be interesting to characterize
better the "initial" turbostratic and the final graphitic crystallites. An interesting point would be to show whether the initial
and final crystallites have similar dimensions. The L_c values
(Table 1) show that the number of stacked layers appears to be somewhat greater in the graphitic component than it is in the turbostratic one. It would be difficult to imagine how the crystallites
could become larger in the process of their turbostratic $\rightarrow$ graphitic
conversion. A more probable explanation is that the L_c values could
not be correctly measured. In particular, no correction was applied
for the strain broadening of the x-ray profiles. Since the strain
broadening is probably larger for the turbostratic structure than
for the graphitic one, the calculated L_c values may be considerably
smaller than the real ones in case of the turbostratic crystallites,
but the difference between calculated and real L_c values should be
less in case of the graphitic component.

A particularly interesting point stems from the experimental
results which point to the conclusion that the crystallites having
a turbostratic structure are suddenly converted into the graphitic
structure. This conversion process is very different from that
occurring in the undoped or B-doped carbons, in which pairs of turbostratic layers are ordered at random. In the SiPC the whole crystallites and not isolated pairs of layers are subjected to a statistical
conversion process.

D. Conclusions

The study of the properties of medium temperature pyrocarbon containing a small amount of Si introduced during the deposition does
not allow a definite conclusion to be made as to the substitutional

solid solubility of Si in carbon, but the experimental evidence suggests that such solutions are formed.

Similar to B, Si improves the structural properties of the carbon up to a concentration supposed to be its solubility limit. This concentration (0.14 weight percent, or 0.06 atomic percent) is much lower than the solubility of B under similar conditions (1 percent), but its effects are considerable. Above the supposed solubility limit SiC is formed. The electronic properties of the Si-containing carbon point to a raising of the Fermi level, which is probably a consequence of the structural improvement.

A specific graphitization behavior, the mechanism of which is still obscure, is caused by presence of Si in the carbon. Two structural components are already formed in the process of carbon formation, and graphitization consists in a conversion of the initially predominating turbostratic component into the graphitic one, at a rate considerably greater than that for the graphitization of pure carbon.

IV. PHOSPHORUS

Phosphorus is an element in the fifth column of the periodical table of elements, immediately below nitrogen, which is next to carbon. Although P and N atoms have the same outer electron structure and form certain stoichiometrically identical compounds, the two elements differ from each other considerably. Nitrogen forms a great number of compounds, the majority of which can be regarded as organic rather than inorganic. In addition, P may be more appropriate for a study of solid solubility in carbon, not only because nitrogen is relatively easily removed from it (although it was found to persist in certain carbons up to 2070 K), but rather because it is difficult to say whether in the latter case we are dealing with solid solubility or with chemical bonding.

A. Structural Properties
The only work on the preparation of P-doped carbon appears to have been done in the author's laboratory [67,68]. Chemical vapor codepo-

sition of carbon and phosphorus was used to prepare carbon containing
a small amount of P. Thermodynamic calculations and experimental
work have shown that at temperatures below 1470 K the deposition becomes extremely slow. On the other hand, the available deposition
apparatus did not permit working at temperatures higher than 2000 K.
Therefore, the temperature range 1490 to 2000 K was used for the
experiments.

Codeposition was effected at 13.3 kPa total gas pressure from a
gas mixture containing CH_4 and PCl_3 (1 to 15 volume percent). In
some experiments an addition of 20 volume percent H_2 was used. The
product of codeposition was similar in appearance to the "pure" pyrocarbon obtained under similar deposition conditions. Chemical analysis has shown that the maximum P concentration that can be retained
in the product decreases with deposition temperature from about 5
weight percent at the lowest temperature to nearly zero at 1800 K.
Microprobe analysis shows an even distribution of P in the product.
The microstructure of the pyrocarbon containing P appears to be the
same as that of the pure PC.

By means of x-ray diffraction no phase other than pyrocarbon
could be detected in the product of codeposition. However, electron
microscopy revealed the existence of two microstructures (Fig. 22),
one of which was similar to that of the undoped PC and the other,
less frequently observed, containing small islands of another phase
dispersed in the mass of PC. From the additional reflections appearing in the electron diffraction patterns it was concluded that the
other phase was cubic red phosphorus with a lattice constant a =
1.131 nm. At higher deposition temperatures, where only a very low
P concentration remains in the PC, no P could be detected by the
electron microprobe, but even in this case single crystals of P
could be sporadically found by electron microscopy. These results
indicated that the product of codeposition was PC with at least part
of the phosphorus present in the form of red phosphorus in the space
between the carbon crystallites.

Further investigations of the PC phase by x-ray diffraction
showed a distinct effect of the presence of P on the interlayer

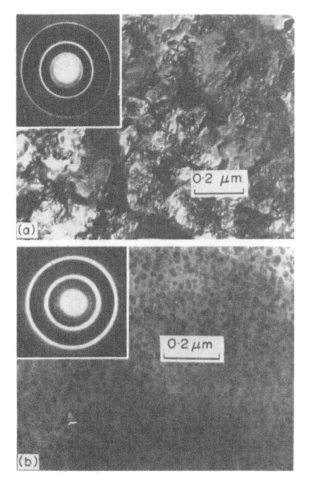

FIG. 22 Typical electron micrographs of phosphorus-containing
pyrocarbon with corresponding diffraction patterns: (a) without
crystallites of phosphorus; (b) with crystallites of phosphorus.
(From Ref. 67.)

spacing and crystallite size, the former increasing and the latter
decreasing with P content (Fig. 23a and c). Since this behavior
indicated the possibility of solid solution formation, it was interesting to compare the results of apparent density measurements with
the values calculated for the hypotheses of substitutional and interstitial solid solutions. Figure 23b shows the results. The points

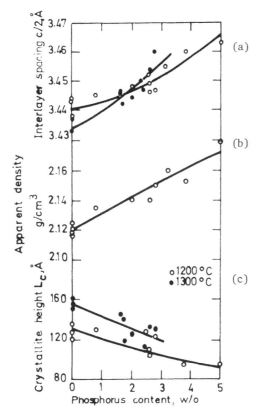

FIG. 23 Dependence of interlayer spacing (a), apparent density (b), and crystallite height (c) of pyrocarbon on its phosphorus content. (From Ref. 67.)

are the experimental values and the line represents the density cal-culated for the hypothesis of substitutional solid solution, using the data from the curve in Fig. 23a and assuming that the amount of unattainable porosity is not affected by presence of P, and that all the P content of the PC is in solid solution. The agreement is fair, although the scatter in the experimental data is obvious. On the other hand, the observed density increase is too small to account for interstitial solution.

The observed increase of interlayer spacing with increasing P content appears to be larger than it could be expected to be on the

TABLE 4   Lattice Strain and Preferred Orientation of Phosphorus-Containing Pyrocarbon (T_D = 1500 K)

Phosphorus content (%)	Lattice strain in C direction ($\eta \times 10^2$)	Preferred orientation parameter, $\Delta\alpha$ (deg)
0	2.15	27
2.0	2.20	31
2.8	2.23	33.5
3.8	2.28	37
5.0	2.51	38

Source: Ref. 67.

basis of the van der Waals radius of P (0.19 nm [6]). An idea that perhaps deserves further attention is that the cause of the large increase in interlayer spacing may be a wrinkling of the carbon layers due to noncoplanar bonds formed by the P atom. The lattice strain is also increased in the presence of P as well as the microhardness of the deposit, while preferred orientation becomes less (Table 4).

The effects of HTT on the PC with the higher P content (3.3 to 4.1 percent) were investigated in the range 1490 to 2300 K in vacuum [68]. The behavior of the P-containing PC on HT is compatible with the picture according to which there are two types of P in the material: the elementary P crystallites present in micropores, and the P atoms present in substitutional sites of the carbon lattice.

The elementary P evaporates upon HT and the resulting vapor exerts pressure on the surrounding carbon crystallites, pushing them apart, thus causing a rapid decrease of apparent density (Fig. 24) and preferred orientation, and preventing crystallite growth. The P content remains almost unchanged due to the closed-pore structure of the pyrocarbon, except at the highest HTT, where it decreases (Table 5).

The other part of the incorporated P present in substitutional sites is responsible for the observed changes of interlayer spacing upon HT (Fig. 25). In the temperature range 1770 to 2070 K this

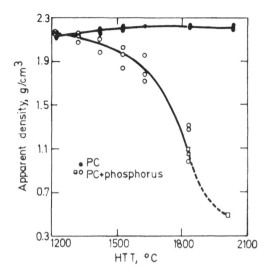

FIG. 24 Apparent density of phosphorus-containing pyrocarbon versus HTT. (From Ref. 68.)

shows qualitatively very similar behavior to the interlayer spacing of pure PC, but remains larger than that, so that the two curves run parallel to each other. Such a behavior indicates that P remains in the carbon lattice during the HT, only leaving it at 2300 K.

TABLE 5 Phosphorus Content, Thickness Ratio, and Apparent Density Ratio of the Heat-Treated Samples

HTT HTT ($^{\circ}C$)	Phosphorus content (%)			Thickness ratio, $t_0/t_T \cong V_0 V_T$			Density ratio, d_T/d_0		
1215	3.3	3.4	4.1				1.00	1.00	1.00
1420	3.1	3.4	4.2				0.92	0.96	0.96
1625	3.5	3.2	4.1				0.80	0.83	0.90
1830	3.0	3.1	3.9	0.47	0.54	0.50		0.49	0.51
2015	2.0	1.9	1.6			0.23			<0.37[a]

[a]Not measurable.
Source: Ref. 68.

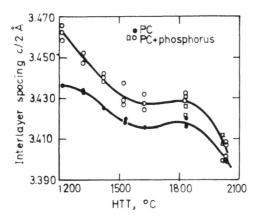

FIG. 25 Interlayer spacing of phosphorus-containing pyrocarbon versus HTT. (From Ref. 68.)

B. Electronic Properties

The only existing data on the electronic properties of phosphorus-containing pyrocarbon refer to the anisotropy of diamagnetic susceptibility, $\Delta\chi$, measured on a number of samples at different temperatures in the range 77 to 700 K, and the longitudinal component of susceptibility, $\chi_\parallel$, measured on a few samples at room temperature [69]. All measurements were performed on the samples used for the determination of structural properties [67,68]. The results are presented in Figs. 26 and 27.

The anisotropy $\Delta\chi$ plotted as a function of P content is shown in Fig. 26. In spite of a scatter in the results, it can be seen that the anisotropy decreases with P content. Figure 27 shows the dependence of total susceptibility, calculated from $\Delta\chi$ and $\chi_\parallel$ values (i.e., $\chi_T = 3\chi_\parallel + \Delta\chi$) versus P content. The susceptibility decreases with P content, in agreement with Fig. 26.

Taking into account the fact that the deposited carbon is very disordered and that it becomes considerably more disordered with increasing P content [67], it would be hardly justifiable to discuss these results in terms of a shift in the Fermi level. However, the decrease of susceptibility occurring with increasing P content indicates an increased trapping of electrons as a consequence of an increasing number of structural defects.

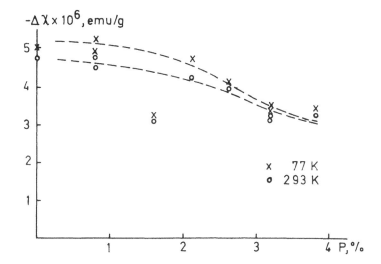

FIG. 26 Dependence of anisotropy of diamagnetic susceptibility on phosphorus content. (From Ref. 69.)

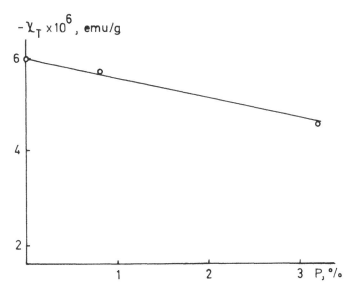

FIG. 27 Dependence of total diamagnetic susceptibility on phosphorus content. (From Ref. 69.)

Since the P atom has a considerably larger size than the C atom
and differs from it in electron structure, it is logical that its
introduction into the carbon lattice should produce distortion. Such
distortion, which may be due to noncoplanar bonds formed by the P
atom, is reflected in an interlayer spacing which increases more than
could be expected from the amount of incorporated P and the size of
the P atom. One possibility would be the suggested wrinkling of the
carbon layers. The distortion is also indicated by the increased
lattice strain and microhardness, as well as by electronic properties,
pointing to an increased trapping of electrons (i.e., an increasing
number of structural defects).

C. Conclusions

Although it would be premature to draw any definite conclusions concerning the substitutional solid solubility of P in carbon from the
existing experimental data, these data indicate such a possibility.
The structural properties of P-containing PC, including their behavior upon HT, may be interpreted assuming, on the basis of the experimental observations, that one part of the P is present in substitutional sites of the carbon lattice, the remaining part being present
as elementary P particles in the spaces between the carbon crystallites. The substitutional P remains stable up to about 2300 K. The
introduction of P leads to an increased disorder of the carbon structure, which is indicated by structural and electronic properties.

V. CONCLUSIONS

1. It can be stated with reasonable certainty that B enters the
trigonal sites of graphite and carbon lattices, substituting for
carbon atoms. The system B-graphite is much the better known. A
part of the B-C phase diagram involving the substitutional solid
solution of B in graphite is proposed. The influence of B on the
structural, electronic, and some other properties of graphite is
quite well established. Less is known about the solubility of B
and its effects on the properties of carbons. There are strong

indications that B enters into the interstitial (interlayer) positions in graphite and carbon lattice, but this point remains to be definitely proved.

2. There are apparently no data about the solid solubility of Si in graphite. As suggested by the existing data involving the structural and electronic properties of Si-containing pyrocarbon, the substitutional solid solubility of Si is much less than that of B. The effects of Si on electronic properties are small. However, Si exerts a strong influence on the graphitization behavior of the carbon, which differs substantially from that of undoped or B-doped carbons. There are no indications of interstitial solid solubility of Si.

3. Only preliminary experiments have been performed concerning the possibility of the solid solution of P in carbon. It is premature to draw conclusions relative to this point, although certain data are in favor of substitutional solid solubility.

REFERENCES

1. J. Millet, J. Millet, and A. Vivares, *J. Chim. Phys. 60*, 553 (1963).

2. A. Marchand, in *Chemistry and Physics of Carbon,* Vol. 7 (P. L. Walker, Jr., ed.), Marcel Dekker, New York, 1971, p. 155.

3. D. B. Fischbach, in *Chemistry and Physics of Carbon*, Vol. 7 (P. L. Walker, Jr., ed), Marcel Dekker, New York, 1971, p. 1.

4. A. Ōya and H. Marsh, *J. Mater. Sci. 17,* 309 (1982).

5. A. Ōya and S. Otani, *Carbon 19*, 391 (1981).

6. F. A. Cotton and G. Wilkinson, *Advanced Inorganic Chemistry,* 2nd ed., Interscience, New York, 1968.

7. J. A. Turnbull, M. S. Stagg, and W. T. Eeles, *Carbon 3,* 387 (1966).

8. C. E. Lowell, *J. Am. Ceram. Soc. 50,* 142 (1967).

9. P. Wagner and J. M. Dickinson, *Carbon 8,* 313 (1970).

10. J. C. Bowman, J. A. Krumhansl, and J. T. Meers, *Proceedings of the SCI Conference on Industrial Carbon and Graphite,* Society of Chemical Industry, London, 1958, p. 52.

11. D. E. Soule, *Proceedings of the Fifth Conference on Carbon,*
     Vol. 1, Pergamon Press, Elmsford, N.Y., 1962, p. 13.

12. J. W. McClure, *Phys. Rev. 119,* 606 (1960).

13. P. Delhaes and A. Marchand, *Carbon 3,* 125 (1965).

14. P. S. Grosewald and P. L. Walker, Jr., *Carbon 6,* 203 (1968).

15. B. J. C. Van der Hoeven, P. H. Keesom, J. W. McClure, and G.
     Wagoner, *Phys. Rev. 152,* 796 (1966).

16. M. E. Preil, D. P. DiVincenzo, R. C. Tatar, and J. E. Fischer,
     Proceedings of the 15th ACS Carbon Conference, University of
     Pennsylvania, Philadelphia, 1981, p. 72.

17. G. Hennig, *J. Chem. Phys. 42,* 1167 (1965).

18. G. Wagoner, private communication to J. W. McClure and Y. Yafet,
     cited in *Proceedings of the Fifth Conference on Carbon,* Vol. 1,
     Pergamon Press, Elmsford, N.Y., 1962, p. 22.

19. R. O. Grisdale, A. C. Pfister, and W. Van Roosbroeck, *Bell Syst.
     Tech. J. 30,* 271 (1951).

20. P. Albert and J. Parisot, *Proceedings of the Third Conference
     on Carbon,* Pergamon Press, Elmsford, N.Y., 1959, p. 467.

21. W. V. Kotlensky, *Carbon 5,* 409 (1967).

22. F. Tombrel, *Rev. Hautes Temp. Refract. 3,* 79 (1966).

23. S. Marinković, Č. Sužnjević, and I. Dežarov, *Carbon 7,* 185 (1969).

24. M. Oberlin and J. Mering, *Carbon 1,* 471 (1964).

25. J. Mering and J. Maire, in *Les Carbones,* Vol. 1 (A. Pacault, ed.),
     Masson, Paris, 1965, p. 129.

26. J. Mering and J. Maire, in *Chemistry and Physics of Carbon,*
     Vol. 6 (P. L. Walker, Jr., ed.), Marcel Dekker, New York, 1970,
     p. 125.

27. S. Marinković, *J. Chim. Phys.,* Spec. No. 84 (1969).

28. W. V. Kotlensky and H. E. Martens, *Carbon 2,* 315 (1964).

29. R. N. Katz and C. P. Gazzara, *J. Mater. Sci. 3,* 61 (1968).

30. C. A. Klein, in *Chemistry and Physics of Carbon,* Vol. 2 (P. L.
     Walker, Jr., ed.), Marcel Dekker, New York, 1966, p. 225.

31. M. C. Robert, M. Oberlin, and J. Mering, in *Chemistry and Physics
     of Carbon,* Vol. 10 (P. L. Walker, Jr. and P. Thrower, eds.),
     Marcel Dekker, New York, 1973, p. 141.

32. R. Gremion, F. Tombrel, J. Maire, and J. Mering, *Compt. Rend.
     260,* 1402 (1965).

33. M. C. Robert and M. Oberlin, *J. Chim. Phys.,* Spec. No. 80 (1969).

34. H. Gasparoux, A. Pacault, and E. Poquet, *Carbon 3,* 65 (1965).

35. P. Delhaes and A. Marchand, *Carbon 3,* 115 (1965).

36. A. Marchand and E. Dupart, *Carbon 5,* 453 (1967).

37. A. Marchand and M. F. Castang-Coutou, *Carbon 9,* 593 (1971).

38. P. Maillard and J. Maire, *Compt. Rend. C267,* 436 (1968).

39. P. A. Thrower and R. M. Mayer, *Physica Status Solidi A 47,* 11 (1978).

40. J. P. Rouchy and J. Mering, *Compt. Rend. C277,* 533 (1973).

41. A. R. Ubbelohde and F. A. Lewis, *Graphite and Its Crystal Compounds,* Clarendon Press, Oxford, 1960.

42. A. V. Emyashev and L. V. Lavrova, in *Konstruktsionnye materialy na osnove grafita,* Metallurgiya, Moscow, 1967, p. 74.

43. N. V. Lavrov, I. I. Chernenkov, and A. V. Emyashev, *Dokl. Akad. Nauk SSSR 184,* 154 (1969).

44. S. Yajima and T. Hirai, *J. Mater. Sci. 4,* 416 (1969).

45. S. Yajima and T. Hirai, *J. Mater. Sci. 4,* 424 (1969).

46. S. Yajima and T. Hirai, *J. Mater. Sci. 4,* 685 (1969).

47. S. Yajima and T. Hirai, *J. Mater. Sci. 4,* 692 (1969).

48. S. Marinković, Č. Sužnjević, I. Dežarov, A. Mihajlović, D. Cerović, and A. Tuković, *Proceedings of the Third SCI Conference on Industrial Carbon and Graphite,* Society of Chemical Industry, London, 1971, p. 102.

49. S. Marinković, Č. Sužnjević, I. Dežarov, A. Mihajlović, and D. Cerović, *Carbon 8,* 283 (1970).

50. A. Tuković and S. Marinković, *J. Mater. Sci. 5,* 543 (1970).

51. Z. Laušević, Č. Sužnjević, and S. Marinković, *High Temp. High Pressures 13,* 221 (1981).

52. S. Ribnikar, *Bull. Soc. Chim. Beograd 44,* 591 (1979).

53. J. C. Rouillon, Thèse, Troisième cycle, Faculté des Sciences, Université de Bordeaux, 1967.

54. J. H. Richardson and E. H. Zehms, *Technical Report TDR-269 (4240-10-3),* Aerospace Corporation, El Segundo, Calif., 1963.

55. W. V. Kotlensky, *Carbon 4,* 209 (1966).

56. D. B. Fischbach and W. V. Kotlensky, *Electrochem. Technol. 4,* 207 (1967).

57. A. S. Schwartz and J. C. Bokros, *Carbon 5,* 325 (1967).

58. S. Dimitrijević, S. Marinković, A. Marchand, S. Flandrois, and J. C. Rouillon, *Proceedings of the Fifth SCI Conference on Industrial Carbon and Graphite,* Society of Chemical Industry, London, 1978, p. 823.

59. S. Dimitrijević, M.Sc. thesis, Faculty of Sciences, University of Belgrade, 1981.

60. S. Dimitrijević and S. Marinković, *Fizika 12*, Suppl. 1, 390 (1980).

61. P. De Kepper, Thèse, Université de Bordeaux I, 1973.

62. E. Poquet, *J. Chim. Phys. 60,* 566 (1963).

63. H. Gasparoux, Thèse, Université de Bordeaux I, 1965.

64. S. Flandrois, unpublished results.

65. A. Marchand, *Ann. Chim. 13,* 469 (1957).

66. E. Stoner, *Proc. R. Soc. A152,* 672 (1935).

67. S. Marinković, Č. Sužnjević, A. Tuković, I. Dežarov, and D. Cerović, *Carbon 11,* 217 (1973).

68. S. Marinković, Č. Sužnjević, A. Tuković, I. Dežarov, and D. Cerović, *Carbon 12,* 57 (1974).

69. A. Marchand, unpublished results.

2

Kinetics of Pyrolytic Carbon Formation

P. A. TESNER

All-Union Research Institute of Natural Gas
Moscow, USSR

I. INTRODUCTION

In the thermal decomposition of hydrocarbons, both gas-phase and
heterogeneous reactions occur simultaneously. The latter take place
on the walls of reactors and result in the formation of different
kinds of pyrolytic carbon (PyC). Furthermore, recently PyC has become of great value as a structural material and as a coating. In
view of the great practical importance of heterogeneous processes,
a large number of studies have been devoted to them and they have
been reviewed in Refs. 1 to 7. However, in spite of this large
amount of work, heterogeneous processes have been studied considerably less than gas-phase ones. This is accounted for chiefly by the
complexity of the processes which are associated with the formation
of a new solid phase and also with a lack of experimental methods
for studying elementary processes on solid surfaces. As a result,
there are few reliable kinetic data even for total, gross reactions
[3], whereas for gas-phase reactions hundreds of elementary reactions
have been studied and kinetic constants for them may be found in hand-
books.

The present chapter generalizes the results obtained for the
last 10 years at the All-Union Research Institute of Natural Gas.
In this sense the chapter is a continuation of an earlier book [3],
but there is a considerable difference between them. The book contained all the quantitative data available in the literature on PyC
formation and their contradictoriness and insufficiency were shown.
A way for the kinetics of this process to be experimentally investigated was outlined. The present chapter gives quantitative results
obtained using this method.

The investigation was initially aimed at obtaining reliable kinetic constants of heterogeneous reactions of thermal decomposition of individual hydrocarbons. For this purpose a simple weight method was developed which permits measurements of these constants in the absence of gas-phase reactions and the inhibiting effect of hydrogen (i.e., at small degrees of hydrocarbon decomposition). As a result, constants of PyC formation rates were obtained for a number of individual hydrocarbons. It was shown that these constants are independent of the properties of the substrate on which PyC is first formed (i.e., they are absolute). In the course of these investigations it was found that the structure of the PyC being formed is closely connected with the kinetics of the heterogeneous reaction. Therefore, structural studies of the PyC film were carried out simultaneously with kinetic measurements.

These studies permitted a determination of the main features of the process and the development of a kinetic model in which PyC formation is considered as a two-stage process: the formation of crystallite nuclei, and crystallite growth as a hexagonal graphite layer in the basal plane. This kinetic model proved to be capable of predicting the presence of a maximum on the rate curve during PyC formation from a binary mixture of hydrocarbons and the structure of the PyC film formed in this case. These predictions have been confirmed experimentally.

Since the proposed kinetic model explains satisfactorily all available experimental results, the present chapter is arranged as follows. First the main features of PyC formation are formulated, principles of the model development are given, and the main equations are derived. Then a list of rate constants for individual hydrocarbons and all experimental results are presented, together with examples of the use of the kinetic model equations.

II. THEORY

The formation of PyC from hydrocarbons of the gaseous phase is a rather complicated process and its mechanism is far from being understood at the molecular level. Generalizations contained in this section present a kinetic model which conforms to the available experimental results and may be considered as only a quantitative phenomenological theory of the process.

A. Physicochemical Features

The main features of PyC formation which were found experimentally are presented below.

1. The initial rate of PyC formation depends on the surface (substrate) nature. As the substrate is being covered with a PyC layer this rate increases or decreases, but for any substrate a thickness of the PyC layer is eventually achieved at which a stationary rate is reached. The thickness of this "transition" layer depends on the nature of the substrate and the hydrocarbon as well as on temperature. All other things being equal, the higher the temperature, the thicker the transition layer, but the shorter the time of its formation.

The presence of a transition layer is brought about by a chemical interaction of the hydrocarbon molecules with the surface mainly to form carbides. As the thickness of the transition layer grows, the content of carbides in it decreases and that of PyC increases. When the stationary rate is reached, PyC begins to be formed which has no substrate elements as impurities. Such a structure of PyC will also be called "stationary." For the following discussion, if there are no special remarks, we mean by the formation rate and the structure of PyC its stationary rate and stationary structure, respectively.

2. The PyC formation rate is first order with respect to the hydrocarbon pressure. For methane this relationship has been proved experimentally to hold over the limits of four orders of magnitude of the pressure variation. For many other hydrocarbons the first order has also been proved, but within smaller pressure variation limits. The apparent deviation from the first order observed for some hydrocarbons arises either due to hydrogen inhibition or to the peculiarities of the process in a layer of a dispersed material at a low partial pressure of the hydrocarbon.

3. The structure of PyC depends on the nature of the hydrocarbon, on temperature, and on the partial pressure of hydrogen, but
it is independent of the substrate properties and partial pressure
of the hydrocarbon and of the properties and partial pressures of
inert diluents.

B. Kinetic Model

The kinetic model of PyC formation is based on experimental features
of PyC growth and the main principles of the theory of the solid
phase growth. PyC [2] is a laminated structure built up of approximately parallel blocks of graphite hexagonal layers. The average
size of these blocks is determined as the size of the region of
coherent scattering of x-rays and they are often called "crystallites." This is only a conditional term because they are not true
crystals of graphite: the hexagonal layers of which they are formed
are arbitrarily turned with respect to their common normal and the
distance between them is more than in a graphite crystal. For our
kinetic considerations we assume the following idealized model of
PyC. We take as a crystallite a part of a graphite hexagonal layer
of one atom thickness and of the length L_a. Further, we suppose
that all crystallites are of the same dimensions, are strictly
parallel to the substrate, and form planar monomolecular layers
spaced at equal distances from one another. Such a structure is
determined by only two parameters: the distance between the layers
and the average size of the crystallite. Since we are considering
the kinetics of layer-by-layer growth of the film, as it occurs from
the substrate upwards, the rate of the process depends on the structure of the upper growing layer, which is determined by the dimension
L_a, whereas the dimension L_c plays no role at all in such a consideration. Therefore, below, when we speak about the PyC structure, it is
mainly the average size of a crystallite L_a which is meant.

As PyC is built up of submicroscopic formations—crystallites—
we assume, as is common when crystallization is considered [8,9],
that its growth is determined by two simultaneous processes: formation of nuclei of crystallites and their growth. As is done in the
case of crystallization, we assume that the size of the crystallites

is determined by a meeting of crystallites growing toward one another. Since we are considering a planar problem, we assume that the nucleus enables a crystallite of one atom thickness to begin growing. We suppose also that the nuclei are formed only at the boundaries of crystallites and that all the nuclei of one molecular layer are formed simultaneously.

The last two assumptions are based on the following considerations. Carbon atoms which form a boundary between two crystallites belong to different crystallites spaced at different distances from one another and chemical bonds between them are not as saturated as those between the atoms forming a hexagonal graphite layer. Consequently, formation of a nucleus for a new crystallite with participation of these boundary atoms is considerably more probable than with carbon atoms of the hexagonal layer, which has three symmetrical bonds.

Naturally, the nuclei are not formed on the surface simultaneously, but continuously. Therefore, crystallites must be of different sizes. Also, the earlier a nucleus is formed, the larger the crystallite that have grown from it will be. However, the results of an x-ray or electron microscopic study of the PyC film structure give only the average size of a crystallite. Therefore, it is assumed in the model that all crystallites are of equal size and the assumption about the simultaneous formation of the nuclei corresponds to the physical pattern of the model because in this case growth with a constant rate results in the formation of crystallites of the same size.

If a crystallite grows in all directions at the same rate, the boundary of growth is circular and its area is $0.25\pi L_a^2$. However, upon the meeting of two crystallites growing toward each other, their growth will continue until a common boundary is formed, which in an ideal case is straight. In this case the shape of the crystallite will be square. Consequently, the average area of the crystallite is between the values $0.25\pi L_a^2$ and L_a^2. Taking into account that structural measurements yield no information on the shape of crystallites, we assume for simplicity the value L_a^2. Then

$$N = \frac{1}{L_a^2} \tag{1}$$

and

$$P = \frac{4L_a N}{2} \tag{2}$$

In the latter expression the 2 in the denominator denotes that each side belongs simultaneously to two crystallites.

From Eqs. (1) and (2) we obtain

$$P = \frac{2}{L_a} \tag{3}$$

If all crystallites in a molecular layer are formed simultaneously, the time of the formation of one molecular layer equals the time of the crystallite growing over the length $L_a/2$. Consequently,

$$W = \frac{L_a}{2\tau} \tag{4}$$

$$U = \frac{N}{P\tau} \tag{5}$$

$$V = \frac{d}{\tau} \tag{6}$$

Hence

$$V = 2d\,(UW)^{1/2} \tag{7}$$

$$L_a = \left(\frac{W}{U}\right)^{1/2} \tag{8}$$

These are the basic equations of the kinetic model which relate the rates of nucleation and crystallite growth to the rate of growth of a PyC film and the average size of its crystallites. Since the latter two values are determined experimentally, we express the rates of crystallite growth and of nucleation, respectively, by the terms

$$W = \frac{VL_a}{2d} \tag{9}$$

$$U = \frac{V}{2dL_a} \tag{10}$$

The independence of the average size of PyC crystallites of the hydrocarbon partial pressure shows that both the rate of nucleation and the rate of growth of crystallites are first order with respect to the hydrocarbon pressure. This almost apparent result may easily be obtained analytically.

We denote the following: m is the order of the nucleation reaction, n is the order of the crystallite growth, and the indices 0 and p are for the values that refer to the atmospheric pressure and the pressure p, respectively. In accordance with Eqs. (7) and (8), and since V depends on the pressure to the first order and L_a is independent of the pressure (i.e., it is zero order with respect to pressure), the following relationships may be written:

$$V_0 p^1 = 2d(U_0 p^m W_0 p^n)^{1/2} \tag{11}$$

$$L_a p^0 = \left(\frac{W_0 p^n}{U_0 p^m}\right)^{1/2} \tag{12}$$

from whence

$$p^1 = (p^m p^n)^{1/2} \tag{13}$$

$$p^0 = \left(\frac{p^n}{p^m}\right)^{1/2} \tag{14}$$

From this it follows that m = n = 1. Consequently, both the formation of nuclei and the growth of crystallites are of first order with respect to pressure.

Let us write the Arrhenius equations for the processes being considered:

$$V = pC_V \exp\left(-\frac{E_V}{RT}\right) \tag{15}$$

$$U = pC_U \exp\left(-\frac{E_U}{RT}\right) \tag{16}$$

$$W = pC_W \exp\left(-\frac{E_W}{RT}\right) \tag{17}$$

$$L_a = C_L \ \exp \ \left(\frac{E_L}{RT}\right) \qquad (18)$$

The exponential temperature dependence of L_a which allows us to express this relationship in the form of the Arrhenius equation (18) illustrates a close connection between the kinetics and the structure and shows that L_a should, in fact, be considered as a kinetic parameter of the process. But naturally, E_L in Eq. (18) bears no physical sense of an activation energy.

Comparison of Eqs. (15) to (18) with Eqs. (7) and (8) allows us to write

$$E_V = \frac{E_U + E_W}{2} \qquad (19)$$

$$E_U = E_V + E_L \qquad (20)$$

$$E_W = E_V - E_L \qquad (21)$$

$$C_U = \frac{C_V}{2dC_L} \qquad (22)$$

$$C_W = \frac{C_V C_L}{2d} \qquad (23)$$

These equations permit us to find constants for the equations for the formation and growth rates of crystallites.

A system of two equations, (7) and (8) or (9) and (10), determines unambiguously a heterogeneous or, more precisely, topochemical process of PyC formation. The availability of two equations causes this solution to differ from the description of a chemical heterogeneous process which does not result in the formation of a new solid phase. The second equation is necessary because the growth kinetics, and the structure of the solid phase being formed, are determined by simultaneous nucleation and growth of crystallites.

Consider now the rate of PyC formation from a binary mixture of hydrocarbons. With regard to the first order of nuclei formation and their growth and assuming the additivity of both processes, we may write, with the use of Eq. (7):

$$V_{UW} = 2d\left[\left[(1 - \alpha)U_1 + U_2\right]\left[(1 - \alpha)W_1 + \alpha W_2\right]\right]^{1/2} \tag{24}$$

This equation (nonlinear with respect to α) shows that the rate of PyC formation in the case of a binary mixture does not obey the additivity law of the rates V_1 and V_2, which is described by the following linear equation:

$$V_V = (1 - \alpha)V_1 + \alpha V_2 \tag{25}$$

Introducing into Eqs. (24) and (25) the symbols K, m, and n, we obtain

$$V_{UW} = V_1\left[\left[1 + \alpha(m - 1)\right]/\left[1 + \alpha(n - 1)\right]\right]^{1/2} \tag{26}$$

$$V_V = V_1\left[1 + \alpha(k - 1)\right] \tag{27}$$

Dividing Eq. (26) by (27), we have

$$\beta = \frac{\left[(1 + \alpha(m - 1))/1 + \alpha(n - 1)\right]^{1/2}}{1 + \alpha(K - 1)} \tag{28}$$

The analysis of this equation shows that at both $\alpha = 0$ and $\alpha = 1$, $\beta = 1$. Consequently, between these values of α the ratio β must go through a maximum.

By equating the first derivative $d\beta/d\alpha$ to zero, we obtain the value α, which corresponds to the maximum value of the ratio β:

$$\alpha = \frac{k - (m + n)/2}{(m - 1)(n - 1) - (k - 1)(n + m - 2)/2} \tag{29}$$

Substitution of Eq. (29) into (28) gives

$$\beta_{max} = \frac{\left[(mn + km - kn - m^2)(mn - km + kn - n^2)\right]^{1/2}}{2(mn - km - kn + k^2)} \tag{30}$$

Since $k \gg 1$, $m \gg 1$, $n \gg 1$, $m \gg k$, and $k \gg n$, we obtain from Eq. (29),

$$\alpha \approx \frac{1}{k} \tag{31}$$

Substitution of Eq. (31) into (28) yields

$$\beta_{max} \approx \left[\left(1 + \frac{m}{k}\right) \left(1 + \frac{n}{k}\right) \right]^{1/2} \tag{32}$$

Now we can find the average size of crystallites of PyC formed from a mixture of two hydrocarbons. According to Eq. (8), we write

$$L_m = \left[\frac{(1 - \alpha)W_1 + \alpha W_2}{(1 - \alpha)U_1 + \alpha U_2} \right]^{1/2} \tag{33}$$

On substitution of Eqs. (9) and (10), we obtain

$$L_m = \frac{(1 - \alpha)L_1 + \alpha L_2 V_2 / V_1}{(1 - \alpha)L_1 + \alpha V_2 / V_1 L_1} \tag{34}$$

With the symbols k, m, and n and making allowance for Eq. (8), we have

$$L_m = \left[\frac{1 + \alpha(n - 1)}{1 + \alpha(m - 1)} \right]^{1/2} \tag{35}$$

The use of the equations obtained in this section and their experimental check are given in the sections to follow.

III. MEASUREMENT METHODS

All kinetic measurements were made by a simple weight method under dynamic conditions [3]. A sample of a "substrate" with the known surface was put into a tubular reactor; in this reactor constant temperature was maintained and the hydrocarbon being studied or a mixture of a hydrocarbon with an inert gas or hydrogen was blown through it. Before entering the reactor all the gases were cleaned of moisture and oxygen traces. A gain in the sample weight during a known time allowed for the determination of the rate of the heterogeneous process in g cm^{-2} s^{-1}. The peculiarity of the technique was the controlled decomposition degree of the hydrocarbon at the reactor outlet and measurements at a low decomposition degree. At the highest temperature under study the degree of decomposition was not allowed to exceed 1 percent. The degree of decomposition in the main measurements at lower temperatures did not exceed tenths and even hundredths

of a percent. Therefore, it was possible to be sure that the results
of measurements referred to PyC formation from molecules of the hydrocarbon introduced into the system and not from a mixture of this
hydrocarbon with the products of its pyrolysis.

In most measurements a piece of cloth made of quartz filaments
(6 μm in diameter, specific surface area 0.26 m^2/g), or channel black
(specific surface area of the order of 100 m^2/g) was used as a "substrate." For smoothing the particles of carbon black the sample was
preliminarily processed with methane at 800°C to provide a weight
gain of 10 to 15 percent. This corresponds to covering the particles
with approximately one or two monolayers of PyC.

At relatively low temperatures the substrate on which the formation of PyC was investigated was put into a reaction tube in a quartz
boat. When working with carbon black, this technique was used in all
cases. But with quartz filaments, in order to lower the decomposition degree at relatively high temperatures, the filaments were put
into a quartz tube with an external diameter slightly less than the
inner diameter of the reaction tube. The reaction mixture was then
blown through the quartz filaments. Such a technique allowed measurements for acetylene up to 1100°C [10]. At still higher temperatures
aluminium oxide rods were used as a substrate. They were suspended
in the reaction tube in such a way that the reaction mixture was
flowing in the annular gap about 0.2 mm thick. This permitted us to
raise the measurement temperature up to 1300°C [11].

Most of the measurements were carried out at atmospheric pressure. When mixtures of a hydrocarbon with an inert gas were used,
concentrations were determined by flow rates which were measured by
rotameters or flow meters. For measurements at pressures below
atmospheric, vacuum was provided by means of a roughing pump. For
methane, measurements were made at pressures up to 1.5 MPa. A steel
tubular apparatus has been developed for these measurements [12].
which permits measurements in a quartz tube. Pressurized methane
was fed simultaneously to the reaction quartz tube and the annular
gap between the quartz tube and the heated steel tube. Methane from

the quartz tube and the annular gap was withdrawn and its pressure was reduced separately. This prevented dilution of the reacting methane by hydrogen formed by its interaction with the steel wall of the reactor.

When working with liquid hydrocarbons they were fed into the reactor by bubbling with an inert gas. The bubbler was thermostatted and the concentration of the hydrocarbon was determined by weighing the bubbler.

The reproducibility of the measurement results for PyC formation rates on quartz filaments and carbon black particles lies within the limits of ±7 percent. The accuracy of activation energies of PyC formation obtained on the basis of these data is ±5 kJ mol^{-1}. The accuracy of rate constants calculated by experimental equations may be evaluated as ±20 percent.

In measurements of the PyC formation rate on metals, the thickness of the plates and wires used exceeded considerably that of the quartz filaments. Therefore, the accuracy of these measurements was appreciably lower than with quartz filaments.

PyC samples for structural studies were obtained on optically polished melted quartz plates which were placed into a reaction tube simultaneously with the substrates used for the kinetic measurements. In order to obtain information on the structure of surface layers of the films no more than 5 nm thick, all samples have been investigated by back diffraction methods (electron beam diameter 200 μm) and back microdiffraction (electron beam diameter 10 μm). For the studies by transmission diffraction, thicker films were separated from the quartz with a weak solution of hydrofluoric acid.

The measurement technique was as follows. The average sizes of the PyC crystallites were determined from line profile diffraction patterns. Line profiles were obtained by photometering the plates. For reflections of the type $(10\bar{1}0)$ and $(11\bar{2}0)$ obtained with the sample plane perpendicular to the electron beam, the Warren formula [13,14] for a two-dimensional lattice was used:

$$L_a = \frac{1.84c}{\Delta R} \tag{36}$$

where c = 2.44 nm·mm is the constant of the instrument and ΔR is the line width at half of the intensity maximum. According to Kurdjumov and Pilankevich [15], the texture of PyC films has a strong effect on the shape of lines of the type < h, k, $\overline{h + k}$, 0 > on electron diffraction patterns and leads to a decrease in their symmetry in the case of two-dimensional structures. Also according to Kurdjumov and Pilankevich [15,16], for such films the average crystallite size was determined by the small angle portion (ΔR^M) of the diffraction line width at a half of the intensity maximum:

$$L_a = \frac{0.52c}{\Delta R^M} \tag{37}$$

At film thicknesses above 50 nm, calculations were made with due regard to the instrumental line width. The latter was determined by electron diffraction patterns of pyrographite films of the appropriate thickness grown on platinum and consisting mainly of crystallites of the size 1 μm.

It should be noted that the sizes of crystallites in PyC films obtained from methane and calculated by transmission and back diffraction patterns differ. In the latter case the average size of crystallites was larger. With regard to the dependence of the crystallite size on the film thickness, it was assumed that the sizes of crystallites in the upper layer were determined by the results obtained from back diffraction patterns.

All measurements were made with an EF-2/4 electron microscope (Carl Zeiss, Jena, GDR), with 65 kV accelerating voltage. The accuracy of measurements of crystallite sizes may be evaluated as ±20 percent.

IV. RATE CONSTANTS OF PYROLYTIC CARBON FORMATION

In this section rate constants of PyC formation from a number of individual hydrocarbons are given. For most of the hydrocarbons studied the constants were obtained with the use of quartz filaments as a substrate. These constants, with coefficients of no more than 2, coincide with the stationary constants obtained on metals. With

regard to the considerably lower accuracy of measurements on metals
and the uncertainty connected with different surface roughness of
the samples, this coincidence shows that the stationary rate is independent of the properties of the substrate. Therefore, the constants
obtained on quartz filaments may be considered as absolute constants.

For a number of hydrocarbons and, specifically, for paraffin
hydrocarbons higher than methane, we failed to obtain absolute constants. Due to high gas-phase instability, measurements of PyC
growth rates on quartz filaments under the conditions of low degree
of decomposition of these hydrocarbons proved to be impossible. As
a result of gas-phase reactions at quite low temperatures, ethylene
is formed; the rate of PyC formation from it is two to three orders
higher than from paraffin hydrocarbons. In this connection measurements for these hydrocarbons were made on carbon black at much lower
temperatures than for methane.

A. Constants on Quartz

Equations for the first-order absolute rate constants of PyC formation for a number of individual hydrocarbons are given in Table 1
and Fig. 1. For methane and acetylene the equations obtained are

TABLE 1 Equations for the First-Order Rate Constants of PyC
Formation $(g\ cm^{-2}\ s^{-1}\ Pa^{-1})$

Hydrocarbon	Temperature limits (°C)	Equation	References
Methane	650–1300	$8.0 \times 10^{-3} \exp(-272,000/RT)$	11, 17, 18
Ethylene	500–700	$7.6 \times 10^{-7} \exp(-155,000/RT)$	19
Propylene	550–650	$4.0 \times 10^{-7} \exp(-151,000/RT)$	20
Butadiene	475–600	$7.5 \times 10^{-7} \exp(-142,000/RT)$	21
Benzene	750–870	$3.0 \times 10^{-3} \exp(-230,000/RT)$	22
Toluene	650–850	$9.9 \times 10^{-2} \exp(-243,000/RT)$	23
Xylene	600–750	$15.5 \times 10^{-2} \exp(-239,000/RT)$	23
Naphthalene	750–850	$6.6 \times 10^{-3} \exp(-222,000/RT)$	24
Anthracene	730–900	$19.4 \times 10^{-3} \exp(-218,000/RT)$	25
Acetylene	550–1100	$17.1 \times 10^{-7} \exp(-138,000/RT)$	10, 26

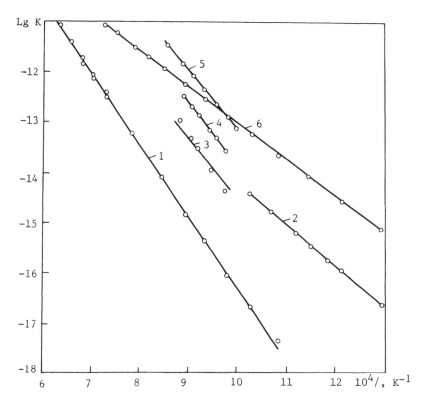

FIG. 1 Logarithms of rate constants of PyC formation from individual hydrocarbons (g cm^{-2} s^{-1} Pa^{-1}): 1, methane; 2, ethylene; 3, benzene; 4, naphtalene; 5, anthracene; 6, acetylene.

true over a very broad temperature range: for methane from 650 to 1300°C, for acetylene from 550 to 1100°C. This result permits extrapolation of these constants to higher as well as to lower temperatures. For other hydrocarbons, the constants were obtained within the temperature limits 100 to 200°C. The high rate of gasphase reaction prevented the broadening of this range. However, it may be expected that the equations obtained are also valid outside the temperature range where the measurements were carried out.

In Tables 2 and 3, constants obtained by the equations from Table 1 are presented. The heavy lines in the tables separate the values extrapolated outside the temperature limits of the measurements. It should be mentioned, however, that these constants can be used for individual hydrocarbons and their mixtures with any

TABLE 2 Rate Constants of PyC Formation from Methane and Acetylene $(g\ cm^{-2}\ s^{-1}\ Pa^{-1})$

Hydrocarbon	Temperature (°C)							
	600	800	1000	1200	1400	1600	1800	2000
Methane	4.24×10^{-19}	4.59×10^{-16}	5.52×10^{-14}	1.81×10^{-12}	2.57×10^{-11}	2.08×10^{-10}	1.12×10^{-9}	4.49×10^{-9}
Acetylene	9.46×10^{-15}	3.27×10^{-13}	3.72×10^{-12}	2.18×10^{-11}	8.40×10^{-11}	2.42×10^{-10}	5.70×10^{-10}	1.15×10^{-9}

TABLE 3 Rate Constants of PyC Formation from Individual Hydrocarbons (g cm^{-2} s^{-1} Pa^{-1})

Hydrocarbon	Temperature (°C)					
	500	600	700	800	900	1000
Ethylene	2.55×10^{-17}	4.04×10^{-16}	3.63×10^{-15}	2.16×10^{-14}		
Propylene	2.50×10^{-17}	3.69×10^{-16}	3.13×10^{-15}	1.78×10^{-14}		
Butadiene	1.90×10^{-16}	2.39×10^{-15}	1.79×10^{-14}	9.16×10^{-14}		
Benzene	8.60×10^{-19}	5.19×10^{-17}	1.35×10^{-15}	1.91×10^{-14}	1.71×10^{-13}	1.09×10^{-12}
Toluene	3.75×10^{-18}	2.85×10^{-16}	8.91×10^{-15}	1.46×10^{-13}	1.49×10^{-12}	1.06×10^{-11}
Xylene	1.10×10^{-17}	7.76×10^{-16}	2.29×10^{-14}	3.59×10^{-13}	3.52×10^{-12}	2.42×10^{-11}
Naphthalene	6.57×10^{-18}	3.44×10^{-16}	7.97×10^{-15}	1.03×10^{-13}	8.58×10^{-13}	5.13×10^{-12}
Anthracene	3.60×10^{-17}	1.75×10^{-15}	3.84×10^{-14}	4.73×10^{-13}	3.80×10^{-12}	2.20×10^{-11}

inert gas (nitrogen, argon, helium, etc.), but in the absence of
hydrogen (i.e., at a low degree of decomposition). Hydrogen has a
strong inhibition effect on PyC formation—different for various
hydrocarbons—and this effect has so far been studied insufficiently
to be reliably taken into account in calculations.

The analysis of the given equations shows that the rate of PyC
formation is strongly dependent on the nature of the hydrocarbon.
As a rule, the higher the absolute value of the rate constant, the
less the activation energy. As a result, at increasing temperatures
the PyC formation rates tend to converge, as is clearly illustrated
in Fig. 1. Acetylene has the highest constants and the lowest ones
refer to methane. The ratio of the constants for these hydrocarbons
at 600°C is 2×10^4 and at 1300°C it is 6.1. A simultaneous solution
of the kinetic equations for methane and acetylene makes it possible
to determine the temperature at which the rates of PyC formation from
these hydrocarbons become equal. This temperature equals 1634°C.

Rate constants for most of the hydrocarbons studied lie between
the values for acetylene and methane. The greatest activation energy
is observed for PyC formation from methane (272 kJ mol^{-1}). Somewhat
lower values are observed for aromatic hydrocarbons (220 to 240 kJ
mol^{-1}) and much smaller values for olefines, diolefines, and acetylene
(140 to 155 kJ mol^{-1}). It should be noted that an experimental value
of the activation energy is the total, gross activation energy of a
complicated process in the course of which carbon-hydrogen bonds are
broken and new carbon-carbon bonds are formed in the graphite layer.
Naturally, the observed activation energy is not in any way connected
with the energy of rupturing some bonds in the hydrocarbon molecule.
It is interesting to note that although aromatic hydrocarbons already
seem to have a ready-made "building block" for a graphite hexagonal
layer, the rate of PyC growth from them is lower than from acetylene
and at low temperatures it is lower than from ethylene. Apparently,
due to the presence of a triple bond, an acetylene molecule has—in
the course of PyC formation—considerable advantages over other
molecular structures investigated. It is also interesting that the

rate of PyC formation from toluene and xylene molecules is higher
than from benzene. The ratio of constants for these hydrocarbons
(at 700°) is

$$K_{C_6H_6} : K_{C_7H_8} : K_{C_8H_{10}} = 1:6.6:17$$

It has been shown by special experiments [23] that in PyC formation from toluene and xylenes, CH_3 groups do not break away from the
benzene molecule, but participate in the formation of a hexagonal
graphite layer.

In a homologous series the PyC formation rate increases with
the hydrocarbon mass. For example, the ratio of rates for a series
benzene:naphthalene:anthracene is 1:5:25.

A large difference in the rates of PyC formation for various
hydrocarbons explains why the experiments in which a low degree of
decomposition of the initial hydrocarbon was not maintained could
not yield reliable kinetic constants.

In fact, for gas-phase pyrolysis of hydrocarbons, on the one
hand, hydrogen appears in the pyrolysis gas and slows down the PyC
formation, while, on the other hand, olefines and aromatics are
formed (and also, at higher temperatures, acetylene); the rates of
PyC growth from these products appreciably exceed those from the
initial hydrocarbons.

Unfortunately, the constants obtained cannot be compared with
the data of other authors since kinetic measurements made in the
absence of gas-phase reactions and hydrogen inhibition are lacking
in the literature. The data for acetylene, which differ only
slightly from the results obtained in Refs. 27 and 28, are an
exception.

B. Constants on Carbon Black

Table 4 presents the equations for the first-order constants obtained
on the surface of carbon black particles. As shown in the table,
these constants were obtained at appreciably lower temperatures than
the constants on quartz (Table 1). For butane it was possible to
obtain only one point at 600°C. At higher temperatures the PyC

TABLE 4 Equations for the First-Order Rate Constants of PyC Formation on Carbon Black (g cm^{-2} s^{-1} Pa^{-1})

Hydrocarbon	Temperature limits (°C)	Equation	Reference
Methane	700-950	1.84×10^{-3} exp(-272,000/RT)	29
Ethane	610-700	4.9×10^{-3} exp(-268,000/RT)	30
Propane	560-610	4.6×10^{-3} exp(-264,000/RT)	20
Butane	600	4.5×10^{-14}	20
Benzene	600-750	18.5×10^{-5} exp(-217,000/RT)	31
Ethylene	500-700	3.5×10^{-7} exp(-155,000/RT)	19
Acetylene	450-600	1.00×10^{-7} exp(-126,000/RT)	26

formation rate in the absence of gas-phase reactions could not be measured. Although the constants obtained on the surface of carbon black particles cannot be considered as absolute ones, they permit evaluation of the relative rates of PyC formation from the first terms of the homologous series of normal hydrocarbons. At 600°C the ratio of constants is

$$K_{CH_4} : K_{C_2H_6} : K_{C_3H_8} : K_{C_4H_{10}} = 1:5.9:9.5:5.7$$

It should be remembered that these constants are expressed in g cm^{-2} s^{-1} Pa^{-1} and in mole units these rates will be respectively smaller. Therefore, it may be stated that, in spite of the considerably greater thermal stability of methane compared to hydrocarbons C_2-C_4, the rate of PyC formation per molecule of methane is only two to three times lower than for these other hydrocarbons. The activation energies of PyC formation for these hydrocarbons are practically the same since they differ only by 4 kJ mol^{-1}, which is within the accuracy limits for this measurement.

Comparison of the results obtained for quartz filaments and carbon black shows the following. For methane and ethylene, activation energies of PyC formation on quartz and on carbon black are the same. For benzene and acetylene the activation energy on carbon black is lower by 13 and 12 kJ mol^{-1}, respectively, and this is

slightly in excess of the measurement error. However, for all hydrocarbons the rate on quartz is several times higher than on carbon
black. This ratio at 700°C is: for methane 4.4, for ethylene 2.2,
and for benzene 3.2; for acetylene it is 3.3 at 600°C. With regard
to a small difference in activation energies obtained on carbon
black and quartz, this ratio is either independent or only very
slightly dependent on temperature.

The observed discrepancy of rate constants considerably exceeds
the experimental errors, but its reasons are at present not under-
standable.

C. Rate Constants of Formation and Growth
   of Crystallites

The rate constants of PyC formation given in the previous sections
allow calculations of only the formation rate for a PyC layer. A
complete description of the process requires another equation to
relate the kinetic parameters of the process with the average sizes
of the PyC crystallites. In the present section such data are presented. Also, kinetic equations for the rates of nuclei formation
and crystallite growth are obtained for methane and acetylene, and
for some other hydrocarbons the rates of these processes are calculated for those temperatures for which the average sizes of crystallites are known.

In Table 5 the available data on average sizes of PyC crystallites for a number of individual hydrocarbons are presented. These

TABLE 5   Average Sizes of PyC Crystallites (nm)

Hydrocarbon	Temperature (°C)						References
	600	700	800	900	1000	1100	
Methane			120	80	60	40	17,18
Ethylene	6.0						19
Benzene			1.5				32
Acetylene	6.0				5.0	4.5	33
Propylene	3.0						34
Butadiene	2.0						34

data refer to PyC films obtained in the conditions under which the
kinetic data were measured (i.e., at low degress of hydrocarbon
decomposition).   Therefore, it may be said with certainty that only
molecules of the initial hydrocarbon—and not the products of the
gas-phase reactions—interacted with the surface to form PyC.

The results given in Table 5 show that the nature of the hydrocarbon has a pronounced effect on the average crystallite size.
Ethylene and acetylene at 600 to 700°C produce crystallites of the
same average size: 6 nm (i.e., 20 times less than methane at 800°C).
At 800°C the average size of crystallites from benzene is only 1.5 nm
(i.e., four times less than those from ethylene).   As temperature increases, the crystallite size decreases, an effect that is especially
noticeable for methane (at 800°C, L_a = 120, and at 1100°C, L_a = 40 nm).

In Fig. 2 the temperature dependence of the average crystallite
size from methane and acetylene are presented in Arrhenius coordinates.
Although the accuracy of the determination of these values is not high,
the points in Fig. 2 fall on straight lines.   This provides an experimental confirmation of the statement that the average size of a crystallite is a kinetic parameter of the process.   The equations obtained
on the basis of this dependence are as follows:

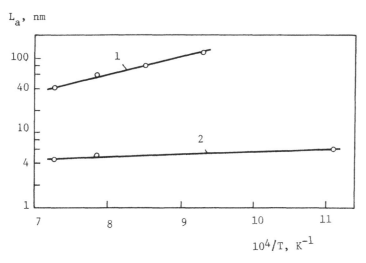

FIG. 2 Average sizes of PyC crystallites: 1, from methane; 2 from
acetylene.

For methane: $L_a = 7.7 \times 10^{-8} \exp\left(\dfrac{45,200}{RT}\right)$ (38)

For acetylene: $L_a = 29 \times 10^{-8} \exp\left(\dfrac{5450}{RT}\right)$ (39)

From these equations, Eqs. (15) to (23), and the constants of PyC growth rates (Table 1), activation energies and equations for the rates of crystallite formation and growth were obtained which are valid within the temperature limits 600 to 1300°C. In these calculations for a conversion of growth constants from g cm^{-2} s^{-1} to cm/s the density of PyC was taken as 2.2 g/cm^3.

Activation energies of formation and growth of crystallites are, respectively: for methane, 317.2 and 226.8 kJ mol^{-1}, and for acetylene, 143.4 and 132.6 kJ mol^{-1}. The equations obtained are:

For methane: $K_U = 7.05 \times 10^{11} \exp\left(-\dfrac{317,200}{RT}\right)$ (40)

$K_W = 4.18 \times 10^{-3} \exp\left(-\dfrac{226,800}{RT}\right)$ (41)

For acetylene: $K_U = 4.0 \times 10^{7} \exp\left(-\dfrac{143,500}{RT}\right)$ (42)

$K_W = 3.36 \times 10^{-6} \exp\left(-\dfrac{132,500}{RT}\right)$ (43)

Table 6 gives the rates of crystallite formation and growth calculated from Eqs. (40) to (43) and extrapolated to 2000°C. The extrapolated values are separated with a heavy line.

It should be noted that there is an appreciable difference between PyC formation from methane and acetylene. For methane the activation energy of the nucleation considerably exceeds that of crystallite growth and, as a result, the average crystallite size rapidly decreases with increasing temperature. For acetylene the difference in the appropriate activation energies is small and the crystallite dimensions are only slightly dependent on temperature.

The constants given in Table 6 show that the rate of nucleation for acetylene is considerably higher than for methane, whereas the rates of crystallite growth differ slightly. Since the activation energies of both processes are appreciably higher for methane than for acetylene, at increasing temperature the ratio of the rates being considered diminishes. In fact, the ratio of nucleation

TABLE 6 Rate Constants of Crystallite Formation (K_U, crystallite cm^{-1} s^{-1} Pa^{-1}) and Growth (K_W, cm s^{-1} Pa^{-1}) from Methane and Acetylene

Hydrocarbon		Temperature (°C)							
		600	800	1000	1200	1400	1600	1800	2000
Methane,	K_U	7.38×10^{-8}	2.55×10^{-4}	6.79×10^{-2}	3.98	8.79×10^{1}	1.00×10^{3}	7.17×10^{3}	3.62×10^{4}
Methane,	K_W	1.12×10^{-16}	3.80×10^{-14}	2.06×10^{-12}	3.79×10^{-11}	3.47×10^{-10}	1.98×10^{-9}	8.06×10^{-9}	2.56×10^{-8}
Acetylene,	K_U	1.04×10^{-1}	4.13×10^{0}	5.17×10^{1}	3.26×10^{2}	1.32×10^{3}	3.98×10^{3}	9.68×10^{3}	2.01×10^{4}
Acetylene,	K_W	3.96×10^{-14}	1.19×10^{-12}	1.23×10^{-11}	6.72×10^{-11}	2.45×10^{-10}	6.78×10^{-10}	1.54×10^{-9}	3.03×10^{-9}

rates for acetylene and methane is 13.7×10^5 at 600°C and at 1200°C
it becomes 82. At the same temperatures the ratios of crystallite
growth rates are 354 and 1.8, respectively.

A simultaneous solution of Eqs. (40) and (42) and (41) and (43)
allows the determination of the temperatures at which the rates of
crystallite formation and growth for acetylene and methane become
equal. These temperatures are 1865 and 1320°C, respectively.

In Table 7 the rates of crystallite formation and growth calculated by Eqs. (9) and (10) are given for a number of individual
hydrocarbons.

D. Calculation of Pyrolytic Carbon Formation
    Rate from a Methane-Acetylene Mixture

For methane and acetylene the constants of both PyC formation and
crystallite formation and growth are known for a sufficiently broad
temperature range. Therefore, it is possible to calculate the rate
of PyC formation for mixtures of these hydrocarbons at 600 to 1300°C.

Table 8 shows the results of the calculation of the rate constants of PyC formation for methane and acetylene and the values of
the parameters K, m, and n. The curves in Fig. 3 are for the value
of the parameter β found by Eq. (28) for α from 10^{-5} to 1 at 600 to
1300°C. The true value of the rate of PyC formation from a methane-
acetylene mixture can be determined by multiplying the additive rate
obtained from Eq. (27) by the value β.

The analysis of the curves in Fig. 3 shows that at low temperatures an appreciable deviation from additivity is observed, especially
at small values of α (e.g., at 600°C, $\beta = 4$ at $\alpha = 6 \times 10^{-4}$). At
temperatures above 1000°C, deviation from additivity decreases, but
it occurs in a sufficiently wide range of α values. For example, at
1300°C the maximum value of $\beta = 1.4$ at $\alpha = 0.1$, but even at $\alpha = 0.5$,
$\beta = 1.2$. Thus a pronounced acceleration of the process is observed,
especially at low temperatures. It is clear that this acceleration
results from a contribution of the nuclei from acetylene whose formation rate considerably exceeds that from methane molecules.

TABLE 7 Rates of PyC Growth and Crystallite Formation and Growth for a Number of Individual Hydrocarbons at Atmospheric Pressure

Hydrocarbon	Temperature (°C)	L_a (nm)	Rate constant (g cm^{-2} s^{-1} Pa^{-1})	V (cm/s)	W (cm/s)	U (crystallite cm^{-1} s^{-1})
Ethylene	600	6.0	4.04×10^{-16}	1.86×10^{-11}	16.7×10^{-11}	4.63×10^{2}
Propylene	600	3.0	3.69×10^{-16}	1.70×10^{-11}	7.61×10^{-11}	8.46×10^{2}
Butadiene	600	2.0	2.39×10^{-15}	1.10×10^{-10}	3.28×10^{-10}	8.21×10^{3}
Benzene	840	1.5	4.81×10^{-14}	2.21×10^{-9}	4.95×10^{-9}	2.20×10^{5}

TABLE 8 Rate Constants of PyC Formation and Crystallite Formation and Growth and Parameters K, m, and n for Methane and Acetylene

Temperature (°C)	K_V (cm s^{-1} Pa^{-1})		K	K_U (crystallite) cm^{-1} s^{-1} Pa^{-1})		m	K_W (cm s^{-1} Pa^{-1})		n
	C_2H_2	CH_4		C_2H_2	CH_4		C_2H_2	CH_4	
600	4.30×10^{-15}	1.93×10^{-19}	2.23×10^4	1.04×10^{-1}	7.38×10^{-8}	142×10^6	3.96×10^{-14}	1.12×10^{-16}	353
700	3.03×10^{-14}	9.09×10^{-18}	3.33×10^3	7.91×10^{-1}	6.59×10^{-6}	120×10^5	2.59×10^{-13}	2.79×10^{-15}	92.8
800	1.47×10^{-13}	2.09×10^{-16}	7.03×10^2	4.13	2.55×10^{-4}	162×10^2	1.19×10^{-12}	3.80×10^{-14}	31.3
900	5.55×10^{-13}	2.80×10^{-15}	1.98×10^2	16.3	5.28×10^{-3}	309×10	4.22×10^{-12}	3.32×10^{-13}	12.7
1000	1.69×10^{-12}	2.51×10^{-14}	67.6	51.7	6.79×10^{-2}	761	1.23×10^{-11}	2.06×10^{-12}	5.97
1100	4.37×10^{-12}	1.63×10^{-13}	26.8	139	6.03×10^{-1}	231	3.06×10^{-11}	9.83×10^{-12}	3.11
1200	9.91×10^{-12}	8.23×10^{-13}	12.0	326	3.98	81.9	6.72×10^{-11}	3.79×10^{-11}	1.77
1300	2.03×10^{-11}	3.37×10^{-12}	6.02	687	20.6	33.3	1.34×10^{-10}	1.23×10^{-10}	1.09

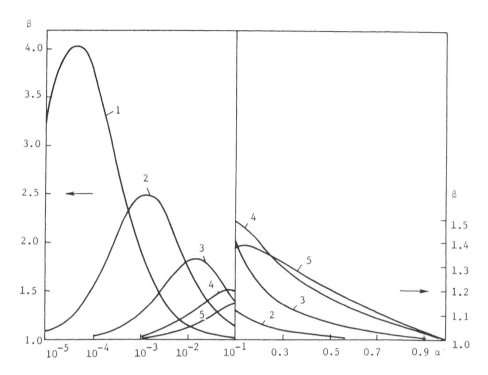

FIG. 3 Calculation of β by Eq. (28) for a methane-acetylene mixture.
Temperature: 1, 600°C; 2, 800°C; 3, 1000°C; 4, 1200°C; 5, 1300°C.

It should be mentioned that the calculation being considered is
valid only for a binary methane-acetylene mixture in the absence of
hydrogen. Therefore, its experimental verification at temperatures
above 1000°C is rather difficult.

V. EXPERIMENTAL CHECK OF THE KINETIC MODEL

In Sec. II.B a possibility was shown of calculating the rate of PyC
formation from binary mixtures of hydrocarbons and of finding the
average size of PyC crystallites for the case of a binary mixture.
The present section deals with work aimed at the experimental verification of these calculations.

A. Rate of Pyrolytic Carbon Formation from
    a Binary Mixture of Hydrocarbons

The use of the kinetic model equations for calculating the rates of
PyC formation from binary mixtures of hydrocarbons showed that this

rate differs considerably from the additive one. An example of such calculation is given in Sec. IV.

In order to check experimentally the calculation of the rate for binary mixtures the rates of PyC formation were measured on quartz filaments for the mixtures methane-acetylene at 600°, methane-ethylene at 700° [35], and methane-benzene at 750°C [32]. The results obtained are shown in Figs. 4 and 5 as a function of the parameters α and β. In the same figures the curves show the results of the rate calculation using Eq. (28).

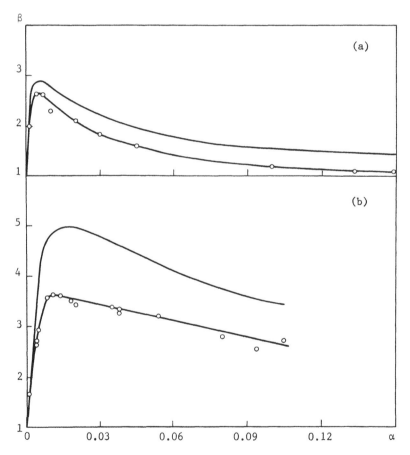

FIG. 4 Comparison of experimental β values (the points) with the calculation using Eq. (28) (the curves): (a) methane-ethylene mixture; (b) methane-benzene mixture.

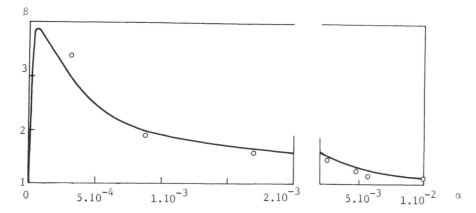

FIG. 5 Comparison of experimental β values (the points) with the
calculation from Eq. (28) (the curve) for a methane-acetylene mixture.

From the analysis of these results it follows that in spite of
some deviation of the experimental points (especially for ethylene
and benzene) they are in satisfactory agreement with the curves calculated using the kinetic model equations. This fact indicates, on
the one hand, the validity of the model and, on the other hand, the
"interchangeability" of the nuclei. In fact, when the equations
were derived to calculate the PyC formation rate from a binary mixture of hydrocarbons, no reservation was made with regard to any
difference in the nuclei formed from different hydrocarbons and,
consequently, it was supposed that the nuclei obtained from one
hydrocarbon are quite suitable for the growth of crystallites from
another hydrocarbon. This supposition is confirmed by satisfactory
agreement of experiment with calculation.

B. Average Size of Pyrolytic Carbon Crystallites
    Formed from a Binary Mixture of Hydrocarbons
For an experimental check of the calculation of average sizes of PyC
crystallites formed from binary mixtures of hydrocarbons, a PyC film
was investigated. This film was obtained from a mixture of methane
(98 volume percent) and ethylene (2 volume percent) [36] on an aluminum oxide rod (1.5 mm in diameter) which was processed in a quartz
tube (12 mm in diameter) at 700°C for 96 h with a mixture flow rate

of 700 cm^3/min. The average size of crystallites in this film was
determined to be 16 ± 4 nm. The average size of a crystallite for
the mixture being investigated was calculated to be 19 nm using Eq.
(35). For this calculation the average size of PyC crystallites
from ethylene was taken as 6 nm (Table 5) and from methane it was
calculated to be 205 nm using Eq. (36).

A comparison of the experimental and calculated values shows a
satisfactory agreement. Thus it may be stated that the structure of
the pyrolytic carbon film predicted on the basis of the kinetic model
has been confirmed experimentally.

VI. HYDROGEN INHIBITION

An inhibiting effect of hydrogen on the PyC formation rate has long
been known [37]. However, only a relatively small amount of quantitative data has been accumulated.

The influence of hydrogen on the PyC formation rate is of interest from two viewpoints: first, for the calculation of the rate of
PyC formation from a mixture of gases containing hydrogen specifically, in various processes since hydrogen is always present in pyrolysis products; and second, for understanding the mechanism of the
process.

The effect of hydrogen inhibition has been studied most completely for methane [2,38,39]. Figure 6 shows the results of measuring the hydrogen retardation effect in a carbon black layer in
vacuum at temperatures from 700 to 1000°C and Fig. 7 presents the
results obtained at atmospheric pressure at 800 to 1000°C with carbon black, quartz filaments, and aluminum oxide rods. In these
figures the rate of PyC formation with respect to the rate in the
absence of hydrogen is plotted as the ordinate.

Comparison of the results obtained in vacuum and at atmospheric
pressure reveals appreciably stronger inhibition in vacuum; for
example, at 800°C in vacuum at a partial hydrogen pressure of 0.1
kPa (Fig. 6) the rate decreases about fivefold, whereas at atmospheric pressure (Fig. 7) the decrease is less than 10 percent.

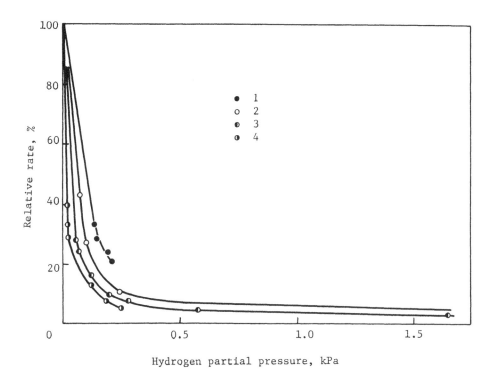

Hydrogen partial pressure, kPa

FIG. 6 Hydrogen inhibition effect on the rate of PyC formation from methane on carbon black in vacuum. Pressure: 7 kPa. Temperature: 1, 700°C, 2, 800°C; 3, 900°C; 4, 1000°C.

It should also be noted that inhibition on the surfaces of carbon black and quartz is different. The curves in Fig. 8 intersect at approximately 35 percent hydrogen in the hydrogen-methane mixture; this fact shows that up to this concentration inhibition on carbon black is stronger than on quartz and above this concentration it is lower. As temperature increases, inhibition in vacuum becomes stronger (Fig. 6) and at atmospheric pressure it decreases (Fig. 7).

The reason for much stronger inhibition in vacuum compared to its effect at atmospheric pressure has become clear only after the deviation from the first order of the reaction in a layer of a dispersed material had been discovered and studied. It was found that this deviation is eliminated by the introduction of hydrogen (see

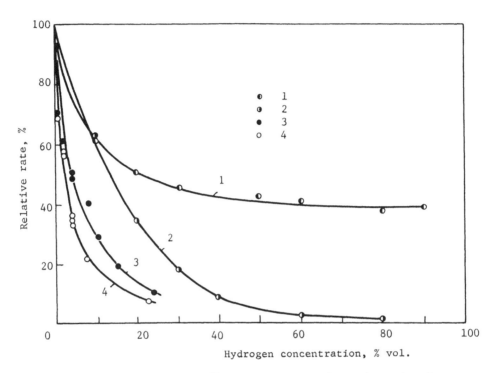

FIG. 7 Hydrogen inhibition effect on the rate of PyC formation from
methane at atmospheric pressure: 1, on aluminum oxide rods, 1200°C;
2, on quartz filaments, 1000°C; 3, on carbon black, 900°C; 4, on
carbon black, 800°C.

Sec. VII.C). On the basis of these results the explanation is as
follows. In vacuum, without hydrogen, PyC is additionally formed
due to the destruction of radicals generated on the surface. The
introduction of hydrogen leads to the elimination of this process.
Therefore, in vacuum the introduction of hydrogen diminishes the
PyC growth rate due to two different mechanisms: first, due to the
elimination of the destruction of radicals, and second, due to inhibition of the molecular growth, whereas at atmospheric pressure it
is due only to the second process. An increase in the inhibition
effect observed in vacuum at increasing temperature is explained by
the increase with temperature of the rate of PyC growth due to the
destruction of radicals.

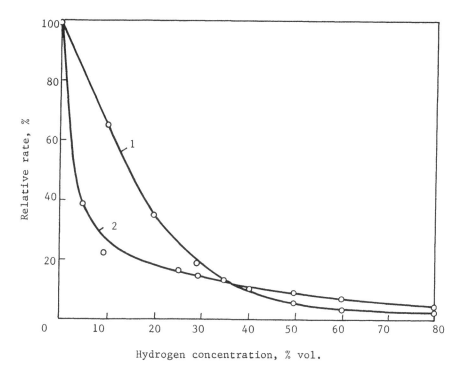

FIG. 8 Hydrogen inhibition effect on the rate of PyC formation from methane at 1000°C on quartz filaments (curve 1) and on carbon black (curve 2) at atmospheric pressure.

In Ref. 40 it was found that the presence of hydrogen in the gaseous phase results not only in a decrease in the rate of the process, but also in a considerable change in the structure of the PyC formed. In Table 9 are given the average sizes of crystallites in PyC films obtained at 1000 and 1200°C from the thermal decomposition of methane from a binary mixture with hydrogen. As shown in Table 9, the higher the hydrogen content in the mixture, the smaller the average crystallite size; for example, at 1000°C in the absence of hydrogen L_a = 60 nm and for 60 percent hydrogen L_a = 6 nm. Thus the hydrogen content in the mixture has an appreciable effect on the structure of PyC being formed.

On the basis of the experimental values of the growth rate and the average crystallite size found for each mixture, the influence

TABLE 9 Effect of Hydrogen on the PyC Structure and the
Formation Rate

Hydrogen concentration (vol. %)	K_V	L_a (nm)	K_L	K_W	K_U
Temperature, 1000°C					
0	1.0	60	1.0	1.0	1.00
10	1.6	40	1.5	2.4	1.07
30	5.4	10	6.0	32	0.90
60	32.0	6	10.0	320	3.20
Temperature, 1200°C					
0	1.0	35	1.0	1.0	1.00
40	2.6	11	3.2	8.3	0.81
90	3.0	7	5.0	15	0.60

of hydrogen may be determined separately for the nuclei formation rate
and the growth rate of crystallites.

The values which refer to methane without hydrogen have a subscript 0 and those for the mixture with hydrogen have a subscript 1
(the index a in the notation L_a is omitted). We denote $K_V = V_0/V_1$,
$K_L = L_0/L_1$, $K_U = U_0/U_1$, and $K_W = W_0/W_1$. Substitution of this notation into Eqs. (7) and (8) gives

$$K_U = \frac{K_V}{K_L} \tag{44}$$

$$K_W = K_V K_L \tag{45}$$

The results of processing the experimental data by these equations
are presented in Table 9. The analysis of these results shows the
following. The value K_U in all cases differs only slightly from
unity. Some scattering in these values may be attributed to the
poor accuracy in the determination of L_a, especially in the region
of small values. Consequently, hydrogen does not practically effect
the rate of nucleation.

The values of K_W increase rapidly with hydrogen concentration:
for example, at 1000°C and 30 percent hydrogen, $K_W = 32$, whereas for
60 percent hydrogen, $K_W = 320$. At 1200°C, inhibition is weaker: at

90 percent hydrogen, $K_W = 15$. Consequently, hydrogen has a pronounced retardation effect on the growth rate of crystallites.

Thus the results obtained show that hydrogen inhibition and
variations in the structure of PyC result from retardation of the
growth rate of crystallites in the basal plane since the rate of
nucleation is practically independent of hydrogen content. It may
be thought that this retardation is brought about by activated adsorption of hydrogen molecules, that is, by chemical interaction of
hydrogen with carbon atoms at a growing step of the carbon layer.

A difference observed in the effect of hydrogen on the nucleation and growth of crystallites provides an additional illustration
of a considerable difference in the mechanisms of these processes.

In Fig. 9 the results for acetylene at temperatures of 600,
1000, and 1250°C are shown [10,38]. Comparison of these results

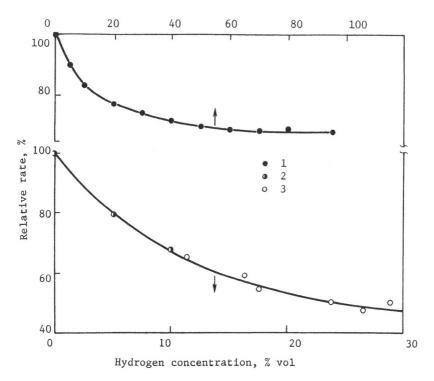

FIG. 9 Hydrogen inhibition effect on the rate of PyC formation from
acetylene at atmospheric pressure: 1, on quartz filaments at 600°C;
2, on quartz filaments at 1000°C; 3, on aluminum oxide rods at 1250°C.

with those for methane shows that, first, hydrogen inhibition in the
case of PyC formation from acetylene is considerably less than in
the case of methane, and second, as temperature increases, inhibition increases, whereas for methane it decreases.

For other hydrocarbons there are fewer data on hydrogen inhibition than for methane and acetylene. Such data for ethane, ethylene,
and benzene [19,30,31] are shown in Figs. 10 and 11. The data for
benzene were obtained both in vacuum and at atmospheric pressure and,
as for methane, they illustrate a considerably stronger effect of
hydrogen inhibition in vacuum. It should be noted that at atmospheric pressure for all the hydrocarbons studied hydrogen inhibition
is considerably weaker than for methane.

Consideration of the experimental curves above, which illustrate
hydrogen inhibition, shows that for all hydrocarbons they look similar and resemble the Langmuir isotherms turned through an angle of
180°. This observation, as well as the well-known fact of hydrogen
adsorption by carbon, was used in several experiments where Langmuir
adsorption by hydrogen was employed, for an explanation and a qualitative description of the hydrogen inhibition effect on PyC formation [3,5].

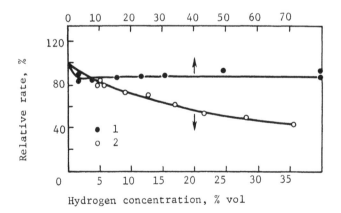

FIG. 10 Hydrogen inhibition effect on the rate of PyC formation
from ethane and ethylene at atmospheric pressure: 1, ethane, 700°C;
2, ethylene, 600°C.

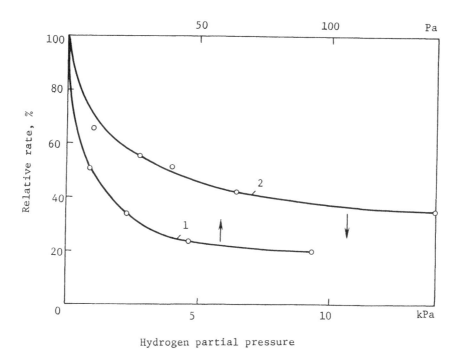

Hydrogen partial pressure

FIG. 11 Hydrogen inhibition effect on the rate of PyC formation
from benzene in vacuum and at atmospheric pressure. 1, Absolute
pressure of benzene 100 Pa. 2, Atmospheric pressure, 700°C; partial
pressure of benzene 13 kPa.

However, the effect of hydrogen on the PyC structure noted
above, and an appreciable difference in the hydrogen inhibition of
PyC formation from methane and acetylene, show that the Langmuir
adsorption of hydrogen cannot explain the mechanism of hydrogen
inhibition. Therefore, the use of the Langmuir equations for its
quantitative description may be just formal.

VII. REACTION ORDER

Most authors believe that PyC formation follows first-order kinetics.
However, in some cases a deviation from the first order is observed
and in some investigations this deviation is used to prove that hydrocarbon adsorption is a limiting stage of the process. In this connection the problem has been studied more comprehensively.

A. First-Order Kinetics

Thermal decomposition of methane was studied over the widest range
of pressures. The results obtained are shown in Fig. 12. Measurements were made at 1050°C on the surfaces of a platinum foil and of
aluminum oxide rods 1 mm in diameter. Platinum foil (35 μm thick)
in pieces 1.5 x 1.5 cm^2 was introduced into a quartz reaction tube
as a pack with the average distance between the pieces being 40 μm.
The aluminum oxide rods were tied in a bundle with platinum wire

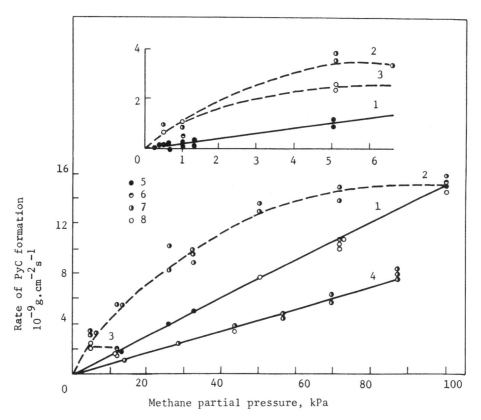

FIG. 12 Rate of PyC formation from methane; temperature 1050°C:
1, the straight line corresponding to a first order reaction;
2, the curve for quartz filaments—density 0.875 g/cm^3; 3 the curve
for quartz filaments—density 0.117 g/cm^3; 4, the straight line for
experiments with the mixture methane-argon-hydrogen; 5, platinum
foil; 6, aluminum oxide rods; 7, quartz filaments—density 0.875
g/cm^3; 8, quartz filaments—density 0.117 g/cm^3.

70 μm in diameter so that the smallest distance between the rods was 70 μm. As shown in Fig. 12, the first order of reaction in the experiments described is preserved over the pressure range 270 Pa to 0.1 MPa.

In Sec. VII.C.2, kinetic measurements are described which were made at 1.5 MPa. These measurements showed that in the pressure range 0.1 to 1.5 MPa first-order kinetics are retained. Thus for methane the first order of the PyC formation reaction has been proved experimentally over the range 270 Pa to 1.5 MPa (i.e., over a pressure range of almost four orders of magnitude). This result allows us to suppose with greater certainty that at both lower and higher pressures the first order of the reaction is preserved.

In Figs. 13 and 14, measurement results for mixtures of ethane, ethylene, acetylene, and benzene with nitrogen are shown. Channel black was used as a substrate in these experiments. These measurements were made over a narrower range of the partial pressure of

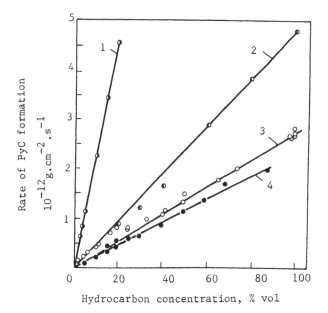

FIG. 13 Rate of PyC formation from ethane, ethylene, and acetylene on carbon black at atmospheric pressure: 1, acetylene, 500°C; 2, ethylene, 600°C; 3, ethane, 700°C; 4, ethane, 700°C—a mixture with nitrogen and hydrogen, hydrogen concentration 10%.

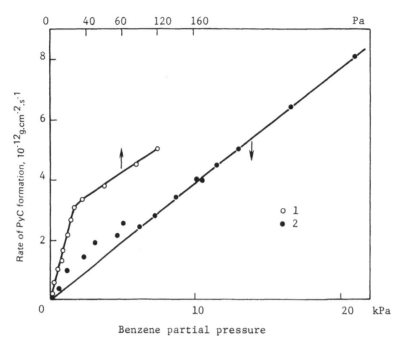

FIG. 14 Rate of PyC formation from benzene mixed with nitrogen on
carbon black: 1, absolute pressure 2 kPa, 800°C; 2, atmospheric
pressure, 700°C.

hydrocarbons than was done for methane, but in all cases the first
order of reaction was found. This result shows that in the course
of PyC formation, hydrocarbon adsorption cannot be the limiting
stage of the process.

     For ethane and benzene at low partial hydrocarbon pressures a
deviation of the rate from first-order kinetics is observed. This
problem is discussed in detail in Sec. VII.C.

B. Influence of Pressure on the Structure
    of Pyrolytic Carbon

Table 10 presents the results of measurements of the average crystallite size for PyC films obtained at the same temperature, but at different pressures of methane. From the analysis of the results listed
in Table 10 it follows that the size of the PyC crystallites is independent of pressure: at 900°C and pressures of 0.1 and 1.5 MPa, the

TABLE 10 Sizes of PyC Crystallites Obtained from Methane

Temperature (°C)	800	900		1000	
Pressure (MPa)	1.5	0.1	1.5	0.026	0.1
L_a (nm)	120	80	80	60	60

average crystallite size is 80 nm, whereas at 1000°C and pressures
of 0.026 and 0.1 MPa, it is 60 nm. This result leads to the conclusion that the rate of nucleation and the rate of crystallite growth,
as well as the growth rate of a PyC film, are first order with respect
to the hydrocarbon pressure.

C. Deviation from First-Order Kinetics
In Sec. VII.A it was shown that PyC formation kinetics for all studied
hydrocarbons are of the first order. In the present section, cases
of apparent deviation from the first order are discussed.

1. *In a Layer of a Dispersed Material*
In the study of the kinetics of PyC formation in a layer of a dispersed material, a peculiar deviation from the first order of reaction is observed. The peculiarity of this deviation is that it is
observed only at a relatively low partial pressure of the hydrocarbon. When a certain pressure is attained, the reaction rate begins
to obey the first-order law. At a partial pressure below that limit
the reaction rate is in excess of the rate corresponding to the first
order observed for a higher pressure. In addition, the limiting
pressure below which a deviation from first-order kinetics is observed depends both on the degree of dispersion and on the packing
density of the dispersed material on which PyC is being formed.

Previously, before these peculiarities of a deviation from
first-order kinetics were found, it was supposed in a number of
papers [5,6] that this deviation is brought about by Langmuir hydrocarbon adsorption. In our work [3] it was assumed that this deviation can be explained by hydrogen inhibition and an incorrect—as
has now become clear—attempt was made to find the values of true

rate constants by extrapolation of the rate to zero hydrogen content. The first work where it was found that the deviation from the first order of the reaction depends on the degree of dispersion of the material [29] was carried out using methane thermal decomposition at 800°C on carbon black and quartz filaments.

In subsequent works the deviation from the first order was studied on quartz filaments of different densities [41] and on diamond powder of different degrees of dispersion.

For experiments with quartz the filaments 40 μm in diameter were used. Two samples of different packings were made. The rate of PyC formation on these samples at 1050°C from methane is demonstrated in Fig. 12. As can be seen from this figure, on the sample of density 0.875 g/cm^3 the deviation from first-order kinetics is observed up to a methane partial pressure of 100 kPa and on the sample of density 0.117 g/cm^3 only up to 10 kPa. For the diamond powder of specific surface area 8.3 m^2/g, the deviation from first order is observed up to 0.1 MPa and for the powder of specific surface area about 2 m^2/g up to 0.05 MPa (Fig. 15).

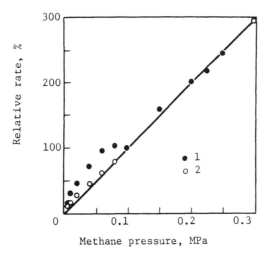

FIG. 15 Rate of PyC formation from methane on diamond powders of different degrees of dispersion. Temperature 850°C: 1, specific surface area 8.3 m^2/g; 2, specific surface area 2.0 m^2/g.

For some hydrocarbons (ethane, benzene), the same as for methane, the deviation from first-order kinetics was found at small partial pressures of the hydrocarbon (Figs. 13 and 14). However, such hydrocarbons as ethylene and acetylene do not exhibit such a deviation (Fig. 13). It should also be noted that the presence of a certain constant concentration of hydrogen in the mixture hydrocarbon-inert gas leads to elimination of this deviation from first-order kinetics. This was shown for methane, ethylene, and benzene (Figs. 12 and 13).

Comparison of the results obtained on various materials leads to the conclusion that the deviation from the reaction first order is not associated with a change in the molecular mechanism of the reaction. In fact, on the surfaces of platinum foil and aluminum oxide rods within the range 270 Pa to 1.5 MPa, no deviation from the first order of reaction is observed. For other materials the pressure range where the deviation from first order is observed is not constant and depends on the packing and degree of dispersion of the material.

These results show that the observed deviation from the reaction order is not connected with a change in the mechanism of the surface reaction of PyC formation, but is determined by a secondary process. This process occurs only when the reaction takes place in a layer of a dispersed material and its intensity depends on the degree of dispersion and on the packing of this material.

In Ref. 29 Tesner et al. proposed that the observed acceleration of the process is caused by the fact that on the surface, together with the direct decomposition of methane molecules to carbon and hydrogen, rupture of these molecules occurs to form radicals $CH_2^{\cdot\cdot}$ or $CH_3^{\cdot}$ without carbon formation on the surface. At a sufficiently high partial pressure of methane these radicals, upon interaction with methane molecules, form stable molecules of ethane or ethylene. At a low partial pressure of methane and in the course of a reaction in a layer of a dispersed material in conditions of the Knudsen regime, the radicals are destroyed on the surface, and this results in the observed acceleration of carbon formation. This suggestion was confirmed by an excess of hydrogen found experimentally in the pyrolysis gas as compared to its stoichiometric quantity, which corresponds to the reaction of methane decomposition into carbon and hydrogen.

A special study [41] was carried out which was aimed at elucidating the reasons for the observed deviation from the first-order reaction; in this study the pyrolysis gas was analyzed for ethane, ethylene, acetylene, and hydrogen content at the same time as the PyC formation rate was determined. In these experiments pure methane was used which contained only 0.07 percent nitrogen and no traces of ethane.

The experiments were carried out on diamond powder (specific surface area 8.3 m^2/g) and on channel black (specific surface area 96 m^2/g). The results obtained are presented in Fig. 16 and in Table 11. The analysis of the results from this table shows that

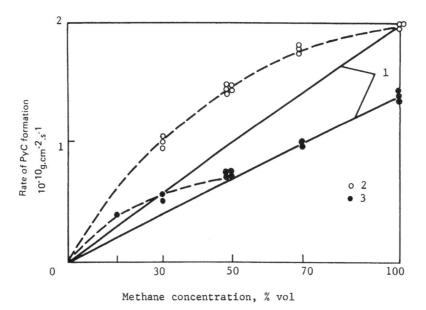

Methane concentration, % vol

FIG. 16 Rates of PyC formation on granulated channel black and diamond powder; temperature 900°C: 1, straight lines corresponding to a first-order reaction; 2, diamond powder; 3, granulated channel black.

TABLE 11 Results of Experiments on Diamond Powder and Granulated Channel Black (Temperature 900°C; Methane Flow Rate 700 cm^3/min)

Experimental conditions	Ratio of surface area to reaction volume (cm^{-1})	Concentration in pyrolysis gas (vol. %)			Decomposition degree (%)
		H_2	C_2H_6	C_2H_4	
In an empty tube	4	0.014	0.006	Traces	0.02
With a diamond powder sample	4×10^3	0.131	0.011	0.005	0.09
With a sample of granulated channel black	4×10^4	0.78	0.031	0.026	0.46

the pyrolysis gas contains no acetylene, but does contain hydrogen and small amounts of ethane and ethylene. The pyrolysis gas in an empty tube contains mainly ethane and only traces of ethylene (less than 0.0005 volume percent). In the tube with a sample the amounts of both ethane and, especially, ethylene in the pyrolysis gas are much greater than those in an empty tube. These results also show that the total degree of decomposition does not exceed 0.46 percent.

The data obtained were processed as follows. It was assumed that increasing concentrations of ethane and ethylene in the experiments with a sample of diamond powder or carbon black are brought about by a contribution of the surface reaction. Therefore, the concentrations of these components due to a heterogeneous reaction were found by subtraction of the appropriate concentrations in the experiments with the sample and in an empty tube. The results are shown in Fig. 17.

On the basis of the gas analysis results, balance calculations were made which showed that the amount of PyC formed corresponds— with a discrepancy of no more than ±2 percent—to the amount of hydrogen formed by the reactions

$$CH_4 = C + 2H_2 \tag{46}$$

$$2CH_4 = C_2H_6 + H_2 \tag{47}$$

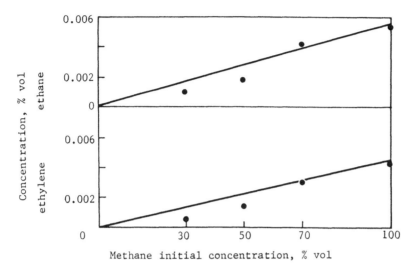

FIG. 17 Concentrations of ethane and ethylene formed on a diamond surface; temperature 900°C.

$$2CH_4 = C_2H_4 + 2H_2 \tag{48}$$

From this one may draw the conclusion that the apparent excess of the actually observed hydrogen concentration over the stoichiometric concentration by reaction (46) is really due to hydrogen formed upon the appearance of ethylene and ethane by reactions (47) and (48).

The quantitative data obtained allowed an evaluation of the surface reaction rates when ethane and ethylene are formed. The results of this calculation are presented in Table 12. As can be seen from the table, the ratio of carbon formed by reaction (46) (i.e., that remaining in the PyC film on the surface) to carbon contained in hydrocarbons C_2 formed as a result of the heterogeneous reaction is: for diamond powder 2.7 and for carbon black 3.2. The rather low accuracy of the analytical results permits their averaging: $(2.7 + 3.2)/2 = 3$. Consequently, from each six methane molecules that break to form PyC on the surface, one methane molecule breaks to form a radical, and this results in the formation of one molecule of ethane or ethylene.

TABLE 12 Calculation of the Rate of Heterogeneous Formation of Ethane and Ethylene (Initial Concentration of Methane 100%)

Sample	Sum of ethane and ethylene concentrations (vol. %)	Mass of carbon in ethane and ethylene (g/experiment)	PyC formation rate, $W_C \times 10^{-10}$ (g cm^{-2} s^{-1})	Formation rate of carbon in C_2 hydrocarbons, $W_{C_2} \times 10^{-10}$ (g cm^{-2} s^{-1})	$\dfrac{W_C}{W_{C_2}}$
Diamond powder	0.0094	3.94×10^{-3}	1.9	0.71	2.7
Granulated channel black	0.051	23×10^{-3}	1.4	0.44	3.2

The experimental results obtained allowed quantitative evaluation of the validity of the assumption about the role of radicals generated by the surface in the formation of additional PyC.

It was assumed that the formation of ethane or ethylene on the surface obeys first-order kinetics. However, the straight lines which express the first-order reaction were drawn only through the points corresponding to methane concentrations of 70 and 100 percent (Fig. 17). This is justified by the fact that at methane concentrations below 70 percent the PyC formation rate deviates appreciably from linearity and, consequently, a decrease in the concentration of C_2 hydrocarbons should be observed due to the destruction of radicals on the surface. In fact, as shown in Fig. 17, the points corresponding to methane concentrations of 30 and 50 percent lie much below the straight lines drawn through the points that correspond to concentrations of 70 and 100 percent.

The observed deviation of the experimental points from the straight lines was used to calculate the deficiency of C_2 hydrocarbons caused by the assumed destruction of radicals. The results of this calculation are shown in Table 13. The last column in this table shows that the deficiency of the carbon mass in C_2 hydrocarbons is only 16 to 20 percent from the increase observed in the amount of PyC formed on the surface. Therefore, the rate increase observed cannot be attributed to the direct destruction of radicals on the surface as was supposed in Ref. 29.

This can be done with the use of the kinetic model of PyC formation. It may be supposed that the destruction of radicals on the surface results in the appearance and growth of new crystallites; this leads to an appropriate general acceleration of the process, although the carbon mass in radicals is only a small portion of the additionally formed PyC. Consequently, a proposal made in Ref. 29 that the observed PyC excess is due to the destruction of radicals generated by the surface is, in principle, correct, but it should not be understood literally. Destruction of radicals makes a small contribution to the increase of PyC mass, but because of the forma-

TABLE 13 Comparison of the Experimentally Observed Gain in the PyC Mass with the Carbon Mass of Radicals Destroyed on the Surface

Initial concentration of methane (vol. %)	Experimentally observed excess of PyC mass over the first-order reaction		Carbon content in C_2 hydrocarbons		Deficiency of carbon mass in C_2 hydrocarbons due to destruction of radicals	
	mg/experiment	%	Calculated by the first order (mg/experiment)	Actually by analysis (mg/experiment)	mg/experiment	%
100	0	0	4.23	4.23	—	—
50	3.3	40	2.12	1.62	0.50	16
30	2.9	52	1.26	0.72	0.54	20

tion of new crystallite nuclei, it leads to an appreciable general
acceleration of the process, and this is observed experimentally.

Shantarovich and Pavlov [42] observed the formation of ethane
(or ethylene) from methane as a result of a heterogeneous reaction
in vacuum. These authors believe that the primary step of methane
cracking is a heterogeneous reaction:

$$CH_4 = \overset{\cdot}{CH}_3 + \overset{\cdot}{H}$$

At homogeneous methane cracking the primary step is, apparently,
also the formation of a methyl radical [43]. Then ethane is formed
at methane interaction with the methyl radical by the reaction

$$\overset{\cdot}{CH}_3 + CH_4 = C_2H_6 + \overset{\cdot}{H}$$

As to ethylene, it may be thought that it is a product of
ethane dehydrogenation by the reaction

$$C_2H_6 = C_2H_4 + H_2$$

But naturally, the mechanisms of ethane and ethylene formation may
be different. This important problem may be solved only by direct
experiments on the identification of radicals generated by the
surface.

It should also be noted that the coincidence of experimental
results obtained in vacuum and at methane dilution by argon shows
that the notion "Knudsen regime" must not be taken literally. In
the experiments with vacuum at a low partial pressure of methane
the Knudsen regime is actually realized and collisions of the radical with the surface become more probable than those with methane
molecules. But as the partial pressure of methane decreases due to
its dilution by argon, the Knudsen regime is not realized but the
same effect of increasing PyC formation rate is observed. The explanation may be as follows. Decomposition of radicals on the
surface is a reaction of the first order, whereas interaction of
radicals with methane molecules is a reaction of the second order.
As pressure decreases, the ratio of the rates of the first- and the

second-order reactions increases and, consequently, it becomes more probable that the radicals will reach the surface.

Regarding the influence of hydrogen, in whose presence the effect of the rate increase disappears, it may naturally be supposed that hydrogen interaction with the radicals results in the formation of methane molecules. It is impossible to say on the basis of the available data if this reaction occurs in the bulk near the surface or on the surface itself, which leads to the cessation of the generation of radicals.

To quantitatively evaluate the acceleration of PyC formation from methane caused by the secondary reaction in a layer of carbon black, the rate was extrapolated to a zero concentration of methane. On the basis of these results the activation energy of the total process (316 kJ mol^{-1}) has been determined and the following equation has been obtained for the rate constant:

$$K = 4.02 \ \exp\left(-\frac{316,000}{RT}\right) \tag{49}$$

Dividing this equation by the equation for the rate constant of PyC formation rate from methane in a carbon black layer in the absence of acceleration (Table 4), we obtain

$$K = 2.18 \times 10^3 \ \exp\left(-\frac{44,000}{RT}\right) \tag{50}$$

This equation shows how many times the rate of the basic process of PyC formation from methane molecules is accelerated by the secondary process of the destruction of radicals. This acceleration increases with temperature: at 600°C it is 5.1 and at 1000°C it equals 34.1.

2. In the Presence of Hydrogen

The inhibiting effect of hydrogen on the rate of PyC formation leads in some cases to an apparent deviation from the first order of the reaction. This deviation is illustrated below by the experimental results on PyC formation on carbon black from methane at pressures from 0.1 to 1.5 MPa [12]. All the experiments were performed at a methane flow rate of 500 cm^3/min and a temperature 800°C in a quartz

tube 12 mm in diameter. In the experiments where no carbon black
sample was introduced into the reaction zone, the amount of hydrogen
in the gas leaving the reaction was: at atmospheric pressure, 0.01
percent, and at 1.5 MPa, 0.04 percent.

Three series of experiments were carried out. Those of the
first series were run on granulated channel black (specific surface
area 96 m^2/g) with a constant sample mass (0.250 g). In the second
series of experiments the masses of the samples of granulated channel black decreased with increasing pressure: from 0.040 g at 0.25
MPa to 0.0015 g at 1.5 MPa (i.e., approximately in inverse proportion to the square of the pressure). This was done on the basis of
the following considerations. The rate of PyC formation is directly
proportional to the hydrocarbon pressure and the diffusion coefficient is inversely proportional to pressure. Therefore, in order to
retain a permanent gradient of hydrogen concentration in the carbon
black layer it was necessary to decrease the sample mass in inverse
proportion to the square of the pressure.

The third series of experiments was conducted on thermal non-
granulated black. Masses of the samples of the thermal black, the
same as in the second series of experiments, decreased with increasing pressure, from 0.150 g at 0.4 MPa to 0.010 g at 1.5 MPa. These
experiments were aimed at decreasing the concentration of hydrogen
formed in the carbon black layer due to the fact that the specific
surface area of thermal black (10 m^2/g) is about 10 times less than
that of channel black.

In addition, at 1.5 MPa, experiments were made on equal-weight
samples (0.0015 g) of granulated channel black with different granule
size. For this purpose granules of sizes 1.0 and 0.1 mm were chosen.

The results of these experiments are shown in Fig. 18. In this
figure the straight line drawn through zero and the experimental
points obtained at atmospheric pressure correspond to a first-order
reaction. In the first series of experiments an appreciable deviation from first-order kinetics and an approximation to zero order at
pressures above 0.5 MPa are observed. In the second series of

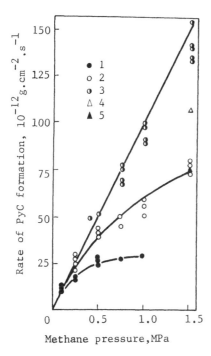

FIG. 18 Rates of PyC formation: 1, first series; 2, second series; 3, third series; 4, on channel black, diameter of granules 1 mm; 5, on channel black, diameter of granules 0.1 mm.

experiments the deviation from first-order kinetics is still appreciable, but less than in the first series. In the third series of experiments a close approach to the straight line is observed.

The observed "straightening" of the curves due to the decreasing weight of the carbon black samples in going from the first to the second series of experiments and from channel to thermal black in the second and third series of experiments is explained by the inhibiting effect of hydrogen; the latter is formed as a result of a reaction in a carbon layer and its removal requires a noticeable gradient in hydrogen concentration in the carbon black layer. A decrease in the sample masses and a decrease in the specific surface of the carbon black lead to a decrease in this gradient.

The influence of the hydrogen gradient is well illustrated by experiments with granules of channel black of different sizes

(Fig. 18). The PyC formation rate for granules 0.1 mm in diameter
is 1.4 times higher than for granules 1 mm in diameter.

In the third series of experiments hydrogen inhibition was
almost completely eliminated due to the use of thin layers of non-
granulated thermal black as the object of investigation. The points
obtained in this series fall almost exactly on the straight line
that corresponds to the first order of the reaction. A small deviation downward from the straight line is caused by slight hydrogen
inhibition, as it is, in principle, impossible to completely eliminate the influence of hydrogen when running a reaction in a layer
of a dispersed material.

The results obtained—besides illustrating the effect of
hydrogen—support the statement that the rate of PyC formation from
methane is first order with respect to methane up to a pressure of
1.5 MPa.

VIII. PYROLYTIC CARBON FORMATION ON CARBON

In the present section the results of measurements of the PyC formation rate on the surface of various carbon structures are given.
The features of the process on the surfaces of carbon black particles,
PyC, and graphite are discussed. In contrast to our discussion in
Sec. VII, it is mainly nonstationary transitions from the initial
carbon structure to the structure of stationary PyC which are considered here.

A. On Carbon Black

PyC formation on the surface of carbon black particles is of interest
from several viewpoints. First, carbon black is an excellent material
for measuring the rate of PyC formation because, due to its large specific surface area, it permits work at relatively low temperatures.
Second, the study of PyC formation on carbon black allows one to
obtain some information on the surface of carbon black particles,
their relative positions, and even to evaluate the specific surface
of carbon black. Moreover, measuring the rate of PyC formation
on the surface of carbon black particles is important for under-

standing a very complicated problem to do with the kinetics of carbon black formation. In the present section these problems are discussed.

1. *Formation of the First Layers*

The formation kinetics of the first PyC layers on the surface of carbon black particles was studied on the following samples of carbon black:

Carbon black	Geometric specific surface area (m^2/g)
Thermal	10
Channel	96
Oxidized channel	110
Acetylene	90
Acetylene, detonation [44]	200

Oxidized channel black was obtained by processing channel black in a nitrogen atmosphere with 1.2 percent oxygen at 800°C. The mass loss in this case was 11 percent. In Fig. 19 are shown the results

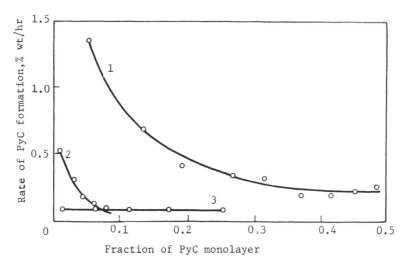

FIG. 19 Rate of PyC formation from benzene on the surface of different kinds of carbon black: 1, channel, preliminarily oxidized; 2, channel; 3, thermal.

obtained for PyC formation from benzene on the surfaces of channel
and thermal black particles. The curves in Fig. 19 reveal a sharp
difference in the behavior of different kinds of carbon black. The
rate of PyC formation on the surface of thermal black remains practically constant up to a coverage of one-fourth of a monolayer,
whereas the rate on the surface of channel black decreases drastically as the surface is being covered with carbon. The initial rate
of PyC formation on the surface of oxidized channel black is, apparently, higher than on the surface of nonprocessed carbon black. The
curves for both samples of channel black are of a similar character.

The change in the structure of the porous surface of carbon
black during its gradual coating with PyC was followed by measuring
the specific surface areas of carbon black samples for different
degrees of coating [45]. Channel black was used in this study.
Coating was performed by the thermal decomposition of benzene or
methane. The specific surface area of the carbon black before and
after processing was measured by adsorption. Figure 20 shows the
dependence of the carbon black specific surface area on its coating
with carbon. From a consideration of this dependence it is seen
that at low degrees of surface coating with carbon (up to a weight
gain of 8 to 9 percent) the specific surface area decreases much
faster than at larger degrees of coating. This fact may be explained
as follows. At first, initially porous particles are smoothed by
carbon being deposited on the surface, and when the surface is coated
with an approximately monomolecular layer of carbon, this smoothing
finishes. A further decrease in the specific surface area is explained by an increase in the total mass of the sample due to the
PyC layer. This conclusion becomes evident if, instead of a true
specific surface area of the sample coated with carbon, a surface
area corresponding to a unit mass of the initial sample is found
(i.e., after a correction is introduced for a gain in the sample
weight upon processing). In Fig. 20 this dependence is shown by a
dashed line. A decrease in the specific surface area referred to
the initial sample is observed only up to a degree of carbon coating
of 8 to 9 percent; at such a gain in weight the specific surface

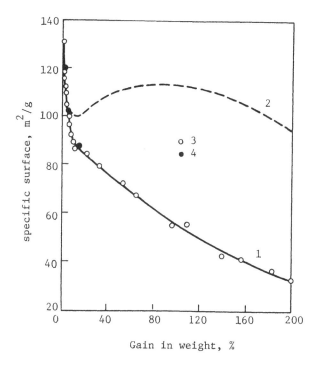

FIG. 20 Specific surface area of channel black: 1, experiment;
2, calculation relative to the initial weight of the sample;
3, samples coated with PyC from benzene; 4, samples coated with
PyC from methane.

area reaches its minimum. Consequently, coating the surface with
an approximately molecular layer of carbon is sufficient to produce
complete smoothing of the original porous surface of the channel
black particles. Further coating of the surface with carbon results
in a certain increase in the surface area of the initial sample
which reaches its maximum at a weight gain of 80 to 100 percent and
then decreases again. It is interesting to note that the coating
of the surface of the particles with a certain carbon mass by the
decomposition of benzene or methane results in the same decrease in
the specific surface.

The rate of PyC formation on the surface of acetylene black
during its coating with carbon does not drop as is the case for PyC
formation on channel black, but increases. In Fig. 21 results are

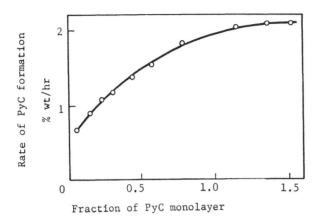

FIG. 21 Rates of PyC formation from benzene on the surface of
acetylene black.

presented which were obtained by coating Shawinigan acetylene carbon
black with PyC by the thermal decomposition of benzene. These results show that the rate increases until the surface is coated with
approximately 1.5 molecular layers of carbon, and afterward the rate
stabilizes. In this case the PyC formation rate is approximately
tripled.

2. Growth of a Collection of Carbon Black Particles
In the preceding section it was shown that the rate of PyC formation
on the surface of various carbon black particles during the deposition of the first molecular layers does not remain constant, but
increases or decreases according to the specific surface structure
of the carbon black particles. However, on coating the surface with
a PyC layer a few atoms thick, the growth rate stabilizes and remains practically constant for a long time. This experimental fact
allowed carbon black to be used as a peculiar "substrate" in measurements of the rate of PyC formation when hydrocarbons are decomposed
on the surface; it also allowed the development of the so-called
kinetic method to measure the specific surface area of carbon black
[46]. Moreover, the coating of porous carbon black particles produces a smoothing of their surfaces and as a result it is possible—
by adsorption measurements of the specific surface area before and

after such a coating—to determine both the degree of porosity of
the carbon black and its geometric surface. A special investigation
was performed [3] which was aimed at a deeper understanding of the
mechanism of PyC growth on a collection of carbon black particles.

Let us consider why the rate of PyC formation on a sample of
carbon black is almost constant after the deposition of a few layers.
If we assume for simplicity that all the particles in this sample of
carbon black are identical isolated spheres which are being coated
with a PyC layer of the same thickness and the density of this layer
equals that of the carbon black particles, then for a factor n increase in the mass of the sample, the diameter of each particle must
increase by $n^{1/3}$ times. The surface area of each particle and, consequently, the total surface area of all particles must increase
$n^{2/3}$ times. For n = 2, $n^{2/3}$ = 1.59, so that when the sample mass
doubles, the total surface area must increase by 60 percent if, as
follows from the experimental results above, the rate is proportional
to the total surface area of the sample. However, experiment shows
that in these conditions the rate increases no more than by 10 to
15 percent. Such a discrepancy lies outside the accuracy limits of
the measurements.

This discrepancy is caused by the specificity of the carbon
black structure [47,48]. Carbon black cannot be considered as a
collection of isolated spherical particles. Separate carbon black
particles are connected to each other in more or less branched
chains. This chainlike structure provides a good explanation for
the growth regularly observed experimentally.

Consecutive experiments on the decomposition of hydrocarbons
on the surface of the same carbon black sample yield an experimental
relationship between the rate of formation of a PyC layer (this rate
is proportional to the absolute value of the surface area accessible
for molecules of the decomposing hydrocarbon) and the total mass of
the sample. We shall find—for different models of relative positions of the particles—an analytical relationship between the total
surface area and the total mass for a given collection of particles
when coated with a PyC layer of given thickness.

Consider the simplest model of the relative positions of the particles. Let all be spherical and of the same diameter, with each particle touching n neighboring particles.

It is clear from elementary geometrical considerations that as the particles are coated with a layer of thickness h (Fig. 22), the surface area of each particle at each point of contact decreases (as compared to the surface area of the total sphere of diameter d + 2h) by the value of the lateral surface area of a sphere segment of height h, and the volume of a particle decreases by the volume of this segment. The surface area f and the volume v of a segment of height h for a sphere d + 2h in diameter are

$$f = \pi(d + 2h)h \tag{51}$$

$$v = \frac{1}{3}\pi h^2[1.5(d + 2h) - h] \tag{52}$$

Thus the surface area F and the mass G of a particle with n points of contact on the surface, as functions of the layer thickness h, are

$$F = \pi(d + 2h)^2 - n\pi h(d + 2h) \tag{53}$$

$$G = \frac{\pi}{6}\delta_1 d^3 + \frac{\pi}{6}\delta_1[(d + 2h)^3 - d^3] - \frac{\pi}{3}n\delta_2(1.5d + 2h)h^2 \tag{54}$$

Here δ_1 and δ_2 are, respectively, the densities of a carbon black particle and of a PyC layer. Division of Eqs. (53) and (54) by the surface area F_0 and the particle mass G_0

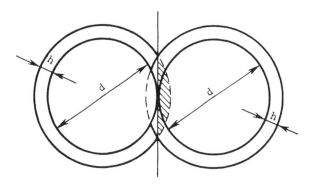

FIG. 22   Scheme of PyC growth on particles in contact.

$$F_0 = \pi d^2 \tag{55}$$

$$G_0 = \frac{\pi}{6}\delta_1 d^3 \tag{56}$$

gives

$$\frac{F}{F_0} = 1 + (4 - n)\ell + 2(2 - n)\ell^2 \tag{57}$$

$$\frac{G}{G_0} = 1 - 2 + 2(1 + 2\ell)^3 - 3n\varepsilon\ell^2 - 4n\varepsilon\ell^3 \tag{58}$$

where

$$\varepsilon = \frac{\delta_2}{\delta_1} \qquad \ell = \frac{h}{d}$$

For the case $\delta_1 = \delta_2$, we have

$$\frac{G}{G_0} = 1 + 6\ell + 3(4 - n)\ell^2 + 4(2 - n)\ell^3 \tag{59}$$

The results of calculations using Eqs. (57) to (59) are plotted in Fig. 23. First, it should be noted that calculations using Eqs. (58) and (59) give very similar results. Consequently in most cases the simpler equation (59) may be used for calculations.

It is most interesting that for n values between 2.5 and 3.5, F/F_0 goes through a maximum which illustrates the possibility of a relatively small deviation of the total surface area from the initial value, even with a considerable change in mass. Thus at n = 3 and a mass change of 500 percent, the maximum deviation of the value F/F_0 is 12.5 percent. For n somewhat higher than 3, an even smaller deviation of the total surface area from its initial value is possible with a three- to fourfold mass increase.

In spite of some conditionality of such considerations, it allows an experimental elucidation of the inner structure of carbon black. The results of such an investigation for channel and acetylene blacks are presented in Fig. 24. In this figure the abscissa is the total mass of the sample expressed as a percentage of the initial value and the ordinate is the observed rate of carbon formation as a percentage of the initial rate. Since the rate of PyC formation is

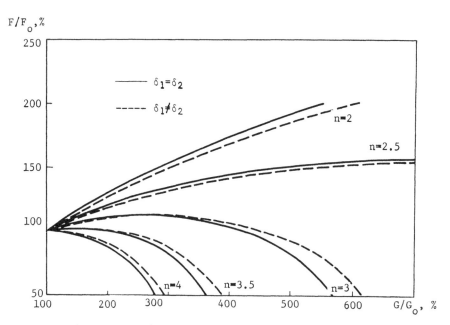

FIG. 23 F/F_0 versus G/G_0.   Calculation: Eqs. (58) and (59).

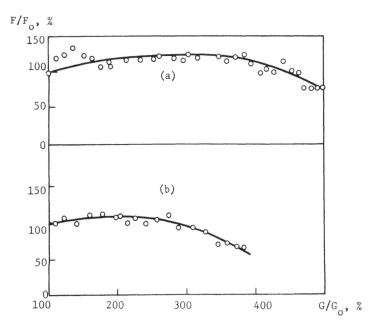

FIG. 24 F/F_0 versus G/G_0 from measurements for (a) acetylene and (b) channel black.

proportional to the surface area, the ratio of surface areas, F/F_0 is plotted instead of the ratio of the rates (analogous to Fig. 23). Thus the experimental coordinates in Fig. 24 correspond to the coordinates in Fig. 23 and the curves in these figures may be compared. This comparison shows that, qualitatively, in both cases the experimental curves correspond closely to the calculated for $n = 3$.

To find the average number of contact points from these experimental results, we must find an analytical expression to relate the average number of contact points n to the maximum value of F/F_0, $(F/F_0)_{max}$.

Equating the first derivative F/F_0 with respect to G/G_0 to zero and performing some transformations, we obtain

$$\frac{h}{d} = \frac{4 - n}{4(n - 2)} \tag{60}$$

Substitution of the value thus obtained in expression (57) gives

$$n = 4\left(\frac{F}{F_0}\right)_{max} - 4\sqrt{\left[\left(\frac{F}{F_0}\right)_{max} - 1\right]\left(\frac{F}{F_0}\right)_{max}} \tag{61}$$

Since $(F/F_0)_{max} \geqslant 1$ and at $h = 4$ the function has no maximum (see Fig. 23), the mathematical possibility of a plus sign before the second (square root) term of the equation has no practical sense. Equation (61) therefore gives the dependence of the average number of contacts on the value of $(F/F_0)_{max}$ which was sought.

In Table 14 the results of the calculation of the parameters describing the growth of PyC on carbon black spheres are presented. The experimental values for $(F/F_0)_{max}$ for channel and acetylene blacks are 123 and 107 percent. The average number of contacts corresponding to these values are: 2.80 for channel black and 3.19 for acetylene black. The calculated values of G/G_0 in this case are, respectively, from Eqs. (58) and (59), for channel black 365 and 391 percent and for acetylene black 207 and 217 percent. Their experimental values are: for channel black 310 percent and for acetylene black 210 percent. Thus the agreement between experiment and calculation is quite satisfactory.

TABLE 14 Calculation Results for Parameters n, h/d, and G/G_0 Corresponding to $(F/F_0)_{max}$

	$(F/F_0)_{max}$											
	105	107	110	112.5	115	120	123	130	140	150
n, Eq. (61)	3.28	3.19	3.07	3.00	2.94	2.84	2.79	2.70	2.61	2.54
h/d, Eq. (60)	0.141	0.170	0.217	0.250	0.282	0.345	0.383	0.464	0.570	0.676
G/G_0, Eq. (58)	188	207	239	263	287	335	365	434	532	639
G/G_0, Eq. (59)	296	217	253	279	304	358	391	473	575	693

Thus the experimental results and elementary geometrical considerations used for their explanation show that for PyC growing on
a collection of carbon black particles the total surface area of
particles remains practically constant up to 300 to 400 percent
weight gains.

Naturally, the calculation above is only approximate. In reality, the particles not only touch one another, but they are already
aggregated into chains and not all the particles are of the same
size as was assumed in the calculation. Moreover, at large values
of h new points of contact appear between the particles which previously did not touch one another and thus the growth rate increases.
Nevertheless, the above model apparently provides a correct qualitative description of the process.

3. *In Conditions of Carbon Black Formation*

The constants of the PyC growth rate obtained from individual hydrocarbons permitted a comparison of the experimental growth rate of
carbon black particles with the calculated one. The growth rate of
carbon black particles under conditions far from those of carbon
black formation were considered in Sec. IV.B where this rate was
shown to be less than the rate calculated.

It was originally supposed [3] that the growth rate of carbon
black particles is independent of their formation. Therefore, one
could think that there is a possibility of calculating the growth
rate of carbon black particles from the constants obtained independently of carbon black formation. This would facilitate the solution of a rather complicated problem about the kinetics and formation
mechanism of carbon black particles. The possibility seems even more
likely when one remembers that the total mass of carbon black particles is determined only by their growth since the mass of their
nuclei is negligibly small. However, a comparison of the experimental growth rate of carbon black particles under conditions of
carbon black formation with calculations from known constants shows
a considerable discrepancy.

Such a comparison was first made by Arefieva and Tesner [49].
In this investigation experiments were run to measure the growth rate

of carbon black particles, and simultaneously to determine the composition of gaseous products along the length of the reaction zone,
under conditions of carbon black formation from the thermal decomposition of methane at 1350 to 1450°C. The growth rate of carbon black
particles was found by measuring the increase in the diameters of
carbon black particles using an electron microscope. It was found
that the growth rate of carbon black particles is 30 to 40 times as
much as the rate calculated from the constants for acetylene and
methane, which are the main components of the pyrolysis gas. The
authors attributed this observation to the accelerating effect of
the nuclei of carbon black particles which are destroyed on the surface. Later it turned out that the constants for acetylene used
were somewhat overestimated [10]. Therefore, the actual ratio of
the experimental rate to the calculated one was even higher.

A comparison of the experimental growth rate of carbon black
particles with the calculations from the constants of the PyC formation rate was later made in three other papers for carbon black
formation from acetylene at pressures below atmospheric [50], for
methane burning in oxygen on a flat burner [51], and in a pilot
reactor for the production of acetylene by the turbulent burning of
methane in oxygen [52]. In these papers an even greater excess of
the experimental rate over the calculated one was found.

Table 15 presents the experimental results given in the papers
cited above, and the growth rates calculated from the constants. In
this calculation it was assumed that the rate of PyC growth from an
acetylene-methane mixture is additive and hydrogen inhibition is
absent. The ratios in the last column show that the actual growth
rate in conditions of carbon black formation exceeds the calculated
value by two to three orders of magnitude.

It should be noted that in some earlier works analogous phenomena were qualitatively observed. Thus in Ref. 53 a considerable
acceleration of methane thermal decomposition was observed in conditions of carbon black formation, and the authors attributed this to
the influence of nuclei of carbon black particles. In Ref. 54 it
was shown that the growth rate of carbon black particles is appre-

TABLE 15 Comparison of the Growth Rate of Carbon Black Particles in Conditions of Carbon Black Formation with the Calculation from Constants

| Temperature (°C) | Partial pressure (10^3 Pa) | | Growth rate (10^{-7} g cm^{-2} s^{-1}) | | | | Ratio: |
| | CH$_4$ | C$_2$H$_2$ | Experiment | Calculation | | | Experiment / Calculation |
				From CH$_4$	From C$_2$H$_2$	Sum	
Thermal decomposition of methane [49]							
1350	46.7	4.39	750	6.50	2.67	9.17	82
1400	27.0	3.45	900	6.84	2.86	9.70	93
1450	21.6	5.65	800	9.67	6.25	15.9	50
Thermal decomposition of acetylene in vacuum [50]							
1400		0.64	520		0.538	0.538	970
1400		1.10	850		0.924	0.924	920
1400		1.50	1020		1.26	1.26	810
1450		0.62	450		0.694	0.694	650
1500		0.60	395		0.882	0.882	450
Methane burning in oxygen on a plane burner [51]							
1080	15.2	3.04	460	0.038	0.244	0.282	1630
1280	28.4	1.42	820	1.61	0.554	2.16	380
Turbulent burning of methane in oxygen [52]							
1410	5.88	7.40	46,000	1.67	6.51	8.18	5600
1400	4.66	8.71	50,000	1.18	7.22	8.40	6000

ciably higher than that of PyC on the walls, and in Ref. 27 it was
found that for carbon black formation the rate of PyC growth and the
activation energy of the process increase considerably compared to
these values in the absence of carbon black formation.

Apparently, in all the cases above, acceleration of the growth
of carbon black particles under conditions of carbon black formation
was observed. This acceleration may be explained by the destruction
of both hydrocarbon radicals and the nuclei of carbon black particles
on the surface. As a result, new nuclei for surface growth appear
on the surface and this causes acceleration of the process.

Thus the nucleation of carbon black particles, which is the
most difficult aspect of the study of carbon black formation, also
has, apparently, an appreciable effect on the growth of particles.
Therefore there is no hope of studying the growth of particles
independent of their formation. Evidently, the formation of nuclei
and the growth of particles are interconnected. Detailed consideration of these problems is beyond the scope of this chapter.

B. On Pyrolytic Carbon

In Refs. 30 and 31, it was found that the initial rate of PyC growth
for the thermal decomposition of hydrocarbons on the surfaces of
carbon black particles coated with a PyC layer depends on the type
of hydrocarbon from which this layer was obtained. Thus, when PyC
was formed from methane on the surface of carbon black particles
coated with a PyC film obtained from ethane, the initial rate was
higher than the stationary one, and for PyC formation from ethane
on a surface coated with PyC obtained from methane it was lower than
the stationary one. A similar picture was observed for PyC obtained
from benzene and methane. In this case the stationary rate was
established upon coating the carbon black particles with one or two
monolayers of PyC. This peculiar "memory" of the surface was attributed to a difference in the structure of PyC formed from different
hydrocarbons.

A special investigation was carried out to study this phenomenon [55]. Two samples of quartz filament were coated with PyC from

the decomposition of methane and ethylene. Then on the PyC obtained
from methane the rate of PyC formation from ethylene was studied and
on the PyC obtained from ethylene, that of PyC from methane. Methane
and ethylene were chosen because the average sizes of the PyC crystallites obtained from these hydrocarbons differ by more than an order
of magnitude (Table 5). To diminish the time required to obtain
stationary PyC films from methane and ethylene at low temperatures,
the quartz filament was preliminarily coated with a 0.1-μm layer of
PyC from methane at 1000°C. On this filament a 10-nm layer of PyC
from methane was obtained at 770°C and a 14-nm layer from ethylene
at 600°C.

Kinetic measurements with ethylene were made at 600°C and with
methane at 770°C. The flow rate of ethylene was 700 cm^3/min and
that of methane 400 cm^3/min. In these conditions the degree of
decomposition of the hydrocarbons did not exceed 0.001 percent.

The experimental results are shown in Fig. 25 as a function of
PyC formation rate on the thickness of the layer. The reaction rate
is expressed with respect to the stationary rate.

As can be seen from Fig. 25, the formation rate of the first
PyC monolayer from ethylene on a film from methane is 2.5 times
below the stationary rate and increases with the formation of new
PyC layers; it reaches the stationary value when 12 monolayers are
formed.

The results for methane indicate the following. The initial
rate of PyC formation exceeds the stationary rate of the reaction
by a factor of 1.6. As the PyC film thickness increases, the rate
decreases and at a thickness of 12 monolayers it becomes stationary.

From the viewpoint of the kinetic model the results obtained
may be naturally explained. Since the rate of nuclei formation is
proportional to the total perimeter of the crystallites, it is inversely proportional to the size of a crystallite. Therefore, when
PyC is formed from ethylene on a substrate of large crystallites
obtained from methane, the initial rate turns out to be less than
the stationary one. As this surface is coated with layers formed
from ethylene, the sizes of crystallites diminish and the formation

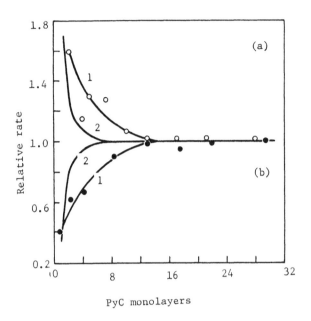

FIG. 25 Rate of PyC formation on quartz filaments coated with PyC.
(a) From ethylene on PyC from methane;(b) from methane on PyC from
ethylene. 1, Experiment; 2, calculation.

rate grows. This goes on until the size of crystallites and the
growth rate reach their stationary values corresponding to ethylene.
When PyC is formed from methane on a substrate of small crystallites,
obtained from ethylene, the initial rate is higher than the stationary
one, but the stationary rate is attained in the same way. Thus the
qualitative explanation of the observed rate variation by a difference in the structure of the PyC obtained from various hydrocarbons
proposed in Refs. 30 and 31 was correct.

Consider qualitatively the results obtained from the kinetic
model. Denote by indices CH_4 and C_2H_4 the parameters of the PyC
substrate and the stationary rates of PyC formation from methane
and ethylene; the index n is used for parameters of the nth layer
of PyC being formed on a substrate from another hydrocarbon. In
so doing we omit the index a in the designation L_a.

Assume that the rate of nucleation and the linear rate of the
growth of crystallites are determined only by the nature of the

hydrocarbon from which a new film is growing, and do not depend on the properties of the substrate. Also assume that they equal the stationary rates. Then, with regard to Eqs. (4) and (5), we have

$$L_n = 2W\tau_n \tag{62}$$

$$N_n = P_{n-1}\tau_n U \tag{63}$$

Elimination of τ_n from these equations and substitution of L_n and L_{n-1} from Eqs. (1) and (3) gives:

$$L_n = \left(L_{n-1}\frac{W}{U}\right)^{1/3} \tag{64}$$

Dividing this equation by Eq. (8), we obtain, for the experiment when the PyC substrate is obtained from ethylene and is being coated with PyC from methane,

$$\frac{L_n}{L_{CH_4}} = \left(\frac{L_{n-1}}{L_{CH_4}}\right)^{1/3} \tag{65}$$

Thus

$$\frac{L_1}{L_{CH_4}} = \left(\frac{L_{C_2H_4}}{L_{CH_4}}\right)^{1/3} \tag{66}$$

$$\frac{L_2}{L_{CH_4}} = \left(\frac{L_1}{L_{CH_4}}\right)^{1/3} = \left(\frac{L_{C_2H_4}}{L_{CH_4}}\right)^{1/3^2} \tag{67}$$

$$L_n = L_{CH_4}\left(\frac{L_{C_2H_4}}{L_{CH_4}}\right)^{1/3^n} \tag{68}$$

Taking Eqs. (7) and (8) into account, we have

$$\frac{L_n}{L_{CH_4}} = \frac{V_{CH_4}}{V_n} \tag{69}$$

Comparison of Eqs. (68) and (69) gives

$$V_n = V_{CH_4}\left(\frac{L_{CH_4}}{L_{C_2H_4}}\right)^{1/3^n} \tag{70}$$

Equations (68) and (70) permit the determination of the average crystallite size and the growth rate for any PyC layer. It is evident that as $n \to \infty$, $V_n \to V_{CH_4}$; and $L_n \to L_{CH_4}$.

These equations are valid only for the case when the substrate is formed by smaller crystallites than the stationary ones formed from the hydrocarbon which build up new layers of PyC. In this case (e.g., for PyC formation from methane on a substrate obtained from ethylene), conditions are followed which form the basis of the kinetic model. If the substrate is formed by larger crystallites and their size decreases as new layers are growing, the conditions of the kinetic model are broken since collisions of crystallites growing toward one another will occur simultaneously in different directions, and this must result in the distortion of their shape. Therefore, in this case the use of these equations is formal and must give a worst approximation.

In Table 16 the results of calculations from Eqs. (68) and (70) are presented. For these calculations it was assumed that $L_{C_2H_4}$ = 6 nm (Table 5) and the value of L_{CH_4} was found by Eq. (38) and equaled 141 nm. The rate V_{CH_4} was found by the kinetic equation (Table 1) for methane for PyC of density 2.2 g/cm^3, and at atmospheric pressure this turned out to equal 8.68×10^{-17} cm/s.

In Fig. 25 the experimental results are compared with calculations. This comparison shows that the shapes of the calculated and experimental curves coincide, but in all cases the calculated curves approach stationary conditions faster than the experimental ones. Thus the calculation shows that approximation of the values L_n and V_n to the stationary ones is already reached in the fifth layer, whereas according to the experimental data, the rate becomes stationary in the twelfth layer. This is due either to the nonvalidity of the assumptions on which Eqs. (68) and (70) were based, or to some phenomena which occur upon a transition from the nonstationary state of PyC growth to the stationary one and which are not accounted for in the kinetic model.

TABLE 16 Calculation of L_n and V_n using Eqs. (68) and (70)

n	Temperature 725°C				Temperature 770°C			
	$\left(\dfrac{L_{C_2H_4}}{L_{CH_4}}\right)^{1/3^n}$	L_n	$\left(\dfrac{L_{CH_4}}{L_{C_2H_4}}\right)^{1/3^n}$	V_n	$\left(\dfrac{L_{C_2H_4}}{L_{CH_4}}\right)^{1/3^n}$	L_n	$\left(\dfrac{L_{CH_4}}{L_{C_2H_4}}\right)^{1/3^n}$	V_n
1	0.322	57.6	3.10	6.48	0.349	49.2	2.86	24.8
2	0.686	122.7	1.46	3.05	0.704	99.3	1.42	12.3
3	0.882	157.8	1.13	2.36	0.889	125.4	1.12	9.72
4	0.959	171.7	1.043	2.18	0.962	135.6	1.040	9.03
5	0.986	176.5	1.014	2.12	0.987	139.0	1.013	8.79
6	0.995	178.1	1.005	2.100	0.996	140.4	1.004	8.72
7	0.998	178.7	1.002	2.094	0.998	140.9	1.001	8.69
8	1.000	179	1.000	2.09	1.000	141	1.000	8.68

C. On Graphite

In the present section the results obtained from measurements of the
initial rate of PyC formation on the basal plane of graphite crystals
upon the thermal decomposition of methane [56] are described. As a
substrate, graphite plates of density 2.266 g/cm^3 were used and with
the parameter c = 0.6710 nm. The average size of the graphite crystallites as determined by optical microscopy is D = 10 μm. The area of
the graphite samples used was 2 to 4 cm^2.

The first experiments were carried out at atmospheric pressure
and a temperature of 900°C. The microanalytical balance used in
this case did not show any weight gain for 15 h (i.e., the total
weight gain was below 2 × 10^{-5} g). In this connection two independent series of experiments were run: at 1.5 MPa and 900°C and at
atmospheric pressure and 1000°C. In so doing it was also found that
in the first experiments no gain in the PyC mass was observed (at
900°C within 6 h and at 1000°C within 2 h).

The results of the experiments are shown in Fig. 26. Here the
ordinate is the first-order rate constant—on a logarithmic scale—
of the PyC formation. The dashed curves connect the experimental
points with the calculated initial rate constants of the formation
of the first PyC monolayer on graphite. In Fig. 26 the values of
the stationary rate constants are plotted for PyC formation at temperatures 900 and 1000°C and the experimental points obtained for
PyC formation on platinum (Fig. 30).

The rate constants for coating of the first PyC monolayer on
graphite have been found in the following way. It was assumed that
the rate of nucleation and the growth rate of crystallites for the
first monolayer on graphite and for the stationary film of PyC were
the same. Then, according to the kinetic model, it may easily be
shown that the ratio of the constant of the stationary rate of PyC
formation to the constant of the growth rate of the first PyC monolayer on graphite is

$$n = \frac{D}{L_a} \tag{71}$$

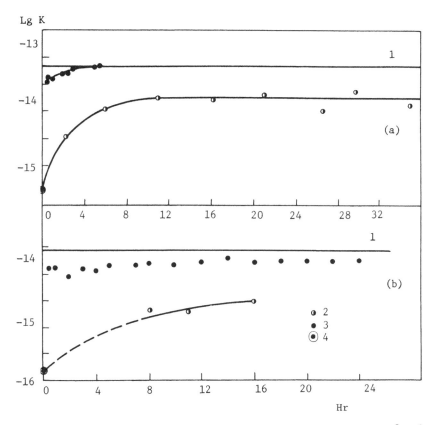

FIG. 26 Logarithms of rate constants of PyC formation (g cm^{-2} s^{-1} Pa^{-1}). A, 1000°C; B, 900°C. 1, Stationary, calculation; 2, experiment; 3, data for platinum (Fig. 30); 4, the first PyC monolayer on graphite, calculation from Eq. (71).

where D is the average size of the graphite crystals. In Table 17 the calculated results for the time of formation of the first PyC monolayer on graphite are presented for temperatures 800 to 1000°C.

The curves in Fig. 26 illustrate an extremely slow increase in the PyC formation rate on graphite. At 1000°C the rate becomes stationary in approximately 10 h after the beginning of processing, whereas on platinum 2 h is sufficient for this. At 900°C in 12 h after the beginning of processing at 1.5 MPa the rate is still 1.5 orders of magnitude below the stationary rate reached on platinum within 6 to 8 h at atmospheric pressure. Simultaneously with the

TABLE 17 Calculation of the Formation Time of the First PyC Monolayer
on Graphite

Temperature (°C)	Rate constant (Table 2) (g cm^{-2} s^{-1} Pa^{-1})	L_a (nm)	n, Eq. (71)	Growth time of the first monolayer, h, at pressure (MPa)		
				0.1	1.5	1.2×10^{-3}
1000	5.52×10^{-14}	60	167	0.617		
900	6.17×10^{-15}	80	125	4.13	0.275	
800	4.59×10^{-16}	120	83	36.9		3080

kinetic measurements, electron microscope measurements were made of
the average size of the crystallites in PyC films formed in the experiments at 1000°C. The following values were obtained: for 15 h
of processing (film thickness 280 nm), L_a = 150 nm, and for 37.5 h
(film thickness 1030 nm), L_a = 60 nm.

These results show that as the PyC film thickness increases,
the crystallite sizes decrease and at the stationary growth rate
they become of size 60 nm, which corresponds to the stationary size
for methane at 1000°C (Table 5).

In fact, these results correspond to those obtained for PyC
formation on PyC and they permit one to explain the observed low
initial growth rate on graphite crystals by the relatively large
sizes of the latter and by the relatively small total perimeter of
the boundaries of these crystals. It should also be noted that, in
contrast to the experiments on PyC, where the stationary rate was
attained when 12 PyC monolayers were formed, on graphite the stationary rate was reached at a thickness of approximately 1000 nm,
which corresponds to approximately 3000 monolayers.

It is interesting to note that Diefendorf showed [57] in 1960
that on the basal plane of the three-dimensionally ordered natural
graphite at 800°C no PyC was observed when processing with methane
at 1.2 kPa over a period of 2 weeks, whereas the quartz reaction
tube became coated with a PyC layer. The results of the present
work explain this observation by the fact that on the basal plane
of the ordered graphite lattice, no crystalline nuclei are formed

and the growth of the first PyC layer begins only at the periphery
of the crystals and therefore proceeds very slowly. A calculation
shows (Table 17) that at 800°C, for a graphite crystal 10 μm across,
and at a methane pressure of 1.2 kPa, the formation of the first PyC
layer must take 3080 h.

True, the reason for such a slow approach to the stationary
rate cannot be explained within the framework of the kinetic model.
Apparently, boundaries between graphite crystals and crystallites
of growing PyC are physically not the same and therefore it is not
correct to assume equal rates of crystallite nucleation. On the
other hand, judging by the curves in Fig. 26, the evaluation of the
formation time of the first PyC monolayer on graphite seems to be
reasonable. Consequently, the formation rate of the first PyC monolayer on the basal plane of graphite and the rate of crystallite
growth for growing PyC are close to one another.

IX. PYROLYTIC CARBON FORMATION ON
     NONCARBON SURFACES

PyC formation on surfaces of substances of different chemical nature
occurs both in laboratories and in technology. In the present section
the results of the investigation of this process are presented. This
investigation was aimed primarily at studying the rate of nonstationary transition processes and measuring the stationary rates on different surfaces.

A. On Quartz

In these measurements a cloth of quartz filaments 6 μm in diameter
was used as a substrate. The results obtained at various temperatures for the thermal decomposition of methane [17,18] are shown in
Figs. 27 and 28. At temperatures of 900°C and higher, the measurements were made at atmospheric pressure and at lower temperatures,
because of the slow rate of PyC formation, at 1.5 MPa.

The analysis of the results obtained shows that in all cases
the curves are alike. At the beginning, a low PyC formation rate is
observed, which then increases and reaches a stationary value. In
this case the lower the temperature, the slower the rate increase.

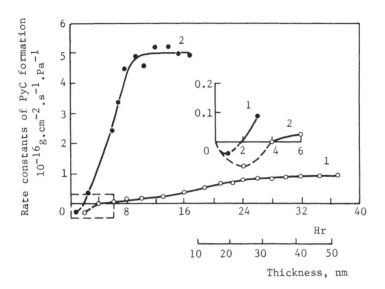

FIG. 27 Rate constants of PyC formation on quartz filaments,
pressure 1.5 MPa: 1, 750°C; 2, 800°C. (From Refs. 17 and 18.)

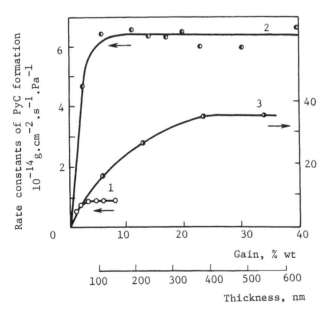

FIG. 28 Rate constants of PyC formation on quartz filaments at
atmospheric pressure: 1, 900°C; 2, 1000°C; 3, 1100°C. (From Refs.
17 and 18.)

At temperatures of 750 and 800°C a weight loss is even observed at
the beginning of the process, in spite of the fact that, visually,
PyC is being formed: the sample is darkening. In this connection
the initial rate constants are negative, which is physically mean-
ingless.

An electron diffraction investigation of PyC thin layers on
quartz indicates the presence of a carbide phase [58]. Therefore,
the weight loss at the beginning is explained by the formation of
silicon carbide.

As the process develops, carbide and PyC are, apparently, formed
simultaneously and the relative amount of carbide decreases with increasing thickness of the layer. As soon as the stationary rate is
attained, carbide formation stops and a stationary film of PyC starts
growing.

In Table 18 results are presented which were obtained by measuring the average size of crystallites and the interplanar distances
for PyC films on quartz as a function of thickness [58], and in Fig.
29 the results of kinetic and structural measurements are compared
for the films obtained from methane at 1000°C. Analysis of the results shows that as the PyC film thickness increases, the crystallite
size also increases, whereas the interplanar distance decreases.

The rate of film growth becomes stationary at a thickness of
200 nm (i.e., when about 600 monolayers of PyC have formed). This
stationary rate corresponds to an average crystallite size (L_a) of
60 nm. As shown in Fig. 29, the average crystallite size becomes

TABLE 18 Average Sizes of PyC Crystallites Formed on Quartz

Time (min)	Film thickness (nm)	L_a (nm)	d_{002} (nm)
10	3	8	0.35
25	8	8	0.35
30	10	15	0.35
200	200	30	0.343
350	460	60	0.336
560	720	60	0.336

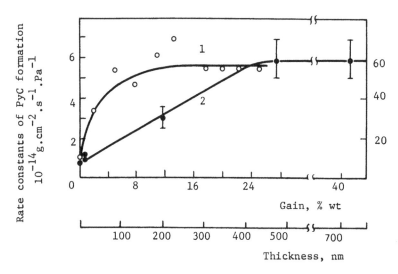

FIG. 29 Rate of PyC formation (1) and crystallite sizes (2) from
methane on the surface of melted quartz at atmospheric pressure.
Temperature 1000°C.

stationary later than the growth rate. This is explained by the
fact that electron microscopic measurements give the average value
L_a over the film thickness, whereas the growth rate depends on the
average size in the upper layer.

The results obtained for the thermal decomposition of ethylene
and acetylene do not differ qualitatively from those obtained for
methane [19,26]. The results obtained at different temperatures
show that the higher the temperature, the thicker the transition
layer, but the less the time required for its formation. This is
well illustrated by Table 19, which gives the conditions in which
the stationary rate is reached for all the hydrocarbons studied.

In concluding this section it should be noted that PyC formation on aluminum oxide and on various refractories with different
proportions of silicon and aluminum oxides is qualitatively of the
same character as on the surface of quartz.

B. On Metals

Carbon formation from the thermal decomposition of hydrocarbons on
the surfaces of various metals has been discussed in a large number

TABLE 19 PyC Formation on Quartz

| Temperature (°C) | Stationary rate sets in | | Reference |
	During the time (h)	At film thickness (nm)	
	Methane, pressure 1.5 MPa		18
750	24	30	
800	8	70	
850	4	120	
900	2	300	
	Methane, atmospheric pressure		17
900	6.7	70	
1000	1.5	150	
1100	1.0	400	
	Ethylene, atmospheric pressure		19
550	20	2	
600	10	4	
660	8	15	
	Acetylene, atmospheric pressure (concentration of acetylene 10 vol. %)		26
550	30	11	
625	7.5	12	
700	5.5	14	
	Propylene, atmospheric pressure		20
650	20	19	

of papers and short reviews [59-63]. In some, the kinetics of the process on various metals was considered [64-67]. Investigations have shown that on most metals the rate of carbon formation is appreciably higher than on carbon since metals have a catalytic effect on the heterogeneous process. For example, on iron, nickel, and cobalt a considerable interaction with methane is observed at temperatures of 500 to 600°C. On carbon surfaces, at such temperatures, the rate of methane decomposition is so small that it may be

measured only when using substances for investigation which have a
large specific surface. Besides, it was found that in contrast to
PyC formation on carbon or quartz, on some metals even at relatively
low temperatures the formation of comparatively large ($\geqslant$10 µm) crystallites of three-dimensionally ordered graphite is observed.

It has been found that the process is accompanied by carbon
diffusion into the metal and that of the metal into a PyC layer,
and also by chemical interaction of the metal with carbon to form
carbides. The process is extremely complicated and its mechanism
depends on the nature of the metal and the hydrocarbon. In spite
of a large number of investigations, it has not been studied ade-
quately.

In the investigations described in this section, formal kinetics
of PyC layer formation on various metals were studied as a function of
time (i.e., on the layer thickness). The purpose of the work was to
find if a stationary rate is achieved at a sufficient thickness of
the PyC layer. The experiments showed that in all cases studied a
stationary rate was not only reached, but it coincided with satisfactory accuracy with the stationary rate found when quartz filaments
were used as a substrate.

Chemical interaction of the substrate with a hydrocarbon, which
determines the presence of a transition layer, was not considered in
this investigation. The initial rate, the thickness of the transition layer, and the time required to attain the stationary rate were
considered as integral parameters of the intensity of this interaction, as a relative measure of the catalytic activity of the substrate with respect to PyC formation.

PyC formation rates were measured by using the same technique
as for the kinetic measurements. But due to the relatively small
geometric surface of the metal samples used, the accuracy of these
measurements was lower and most of the measurements were carried
out only at 1000°C.

1. On Platinum

For measurements of the PyC formation rate on platinum [68], the same
pack of platinum foil leaves (total area 400 cm^2) was used as for

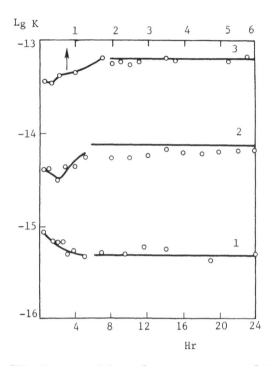

FIG. 30 Logarithms of rate constants of PyC formation from methane
on platinum. Straight lines 1, 2, 3: constants of the stationary
rate at 800°C, 900°C, 1000°C; points, experiments at the same temperatures.

measuring the reaction order (see Sec. VII). The results obtained
are shown in Fig. 30, from which it can be seen that, as for quartz,
after a certain nonstationary region the process rate becomes stationary and differs only slightly from the stationary rate found
for quartz. In contrast to the results obtained on quartz, however,
the initial rate of the process differs rather insignificantly from
the stationary rate: at 900 and 1000°C it is below and at 800°C it
is above the stationary rate. Structural investigations of nonstationary PyC films on platinum showed that, together with fine crystalline PyC, areas of three-dimensionally ordered graphite of the
size 10 to 15 μm are formed. This result agrees with that obtained
for iron, nickel, and cobalt [59]. However, when a stationary rate
is reached, the PyC film is sufficiently uniform and at 1000°C the

average size of crystallites in this film is 60 nm, which agrees
with the results obtained for a stationary rate on quartz.

2. *On High-Melting-Point Metals and*
    Metals of Group VIII

PyC formation on the surfaces of high-melting-point metals and metals
of group VIII differs from that observed on quartz and on platinum.
For all the metals studied [69,70] a high initial rate is observed,
followed by a gradual decrease, and finally a stationary rate is
approached. The experimental results are shown in Fig. 31, and they
show an appreciable individual difference in the values of the initial
rate recorded, in the time required to attain the stationary rate, and
in the thickness of the transition layer. It is clear that each of
these three values characterizes catalytic activity of the metal with
respect to PyC formation. However, because of a rapid decrease of
the initial rate, its reliable measurement presents considerable difficulties. Table 20 gives the ratio of the greatest initial rate to

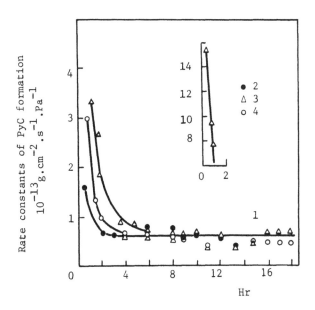

FIG. 31 Rate constants of PyC formation from methane on various
metals at 1000°C: 1, constants of the stationary rate; 2, tungsten-
rhenium (rhenium 20 percent); 3, nickel; 4, molybdenum.

TABLE 20   PyC Formation on Metals at 1000°C and Atmospheric Pressure

NN	Metal	Initial rate / Stationary rate	Stationary rate sets in	
			During the time (h)	At film thickness (μm)
1	Carbon steel	150	14[a]	55
2	Chromium-nickel steel	200	10	40
3	Cobalt	80	16	16
4	Nickel	26	4	3.0
5	Molybdenum	5	2	0.9
6	Tungsten-rhenium (20% rhenium)	3	2	0.4
7	Platinum	1	2	0.2

[a]Stationary rate has not set in.

the stationary one, the time of attaining the stationary rate, and
the transition layer thickness.   It should be noted that as temperature increases, the time required to attain the stationary rate decreases, and the thickness of the transition layer increases.

On the basis of the data listed in Table 20, all metals investigated may be arranged into the following series by their activity
with respect to the process being considered: carbon steel, chromium-
nickel steel, cobalt, nickel, molybdenum, tungsten-rhenium, platinum.

One's attention is drawn to the low activity of platinum and the
high activity of metals of group VIII.   It is interesting to note
that the catalytic activity of steel depends considerably on its
composition.   Thus in chromium-nickel steel the stationary rate is
attained within 10 h at a film thickness of 40 μm, whereas in carbon
steel the stationary rate is not achieved even after 14 h at a film
thickness of 55 μm (Table 20).

X.   DISCUSSION

The results presented above are important mainly because the absolute
constants of the PyC formation rate have been obtained for a number

of individual hydrocarbons and a kinetic model has been constructed
which has been confirmed by experimental data currently available.
Attention should be paid to the fact that the rate of the topochemical process of PyC formation was expressed by simple kinetic equations of a first-order heterogeneous process and the temperature
dependence of the rate was expressed by the Arrhenius equation over
a broad temperature range.

An unambiguous description of the kinetics of the process requires two equations: one for the rate of nuclei formation and the
other for the growth rate of crystallites. This is determined by
the fact that the problem is solved with the use of the theory of
the formation of a new phase. The use of the ideas of the theory
of the formation of the new phase, besides some general considerations, is based on the experimentally found close connection between
the growth kinetics and the structure of the PyC layer being formed.
However, the introduction of the ideas of the foregoing theory has
not changed the form of the kinetic equations. They do not differ
at all from conventional equations which describe the kinetics of a
first-order heterogeneous process. This may, apparently, be explained
by the fact that both the nucleation and the growth of crystallites
are chemical processes. Therefore, they are not limited by such physical parameters of the growth of the solid phase as supersaturation
and the size of a minimum nucleus [71]. If these parameters played
a limiting role, no first-order reaction would be observed either for
the total process or, separately, for nucleation and growth of crystallites, as well as the Arrhenius dependence over a broad range of
temperature variation. This is an important feature of the chemical
crystallization of carbon from a gaseous phase and is the major difference from such processes as the chemical epitaxial crystallization
of silicon or germanium [72]. The two-stage mechanism of PyC formation, when nuclei are formed only on the boundaries of crystallites
and the activation energies of formation and growth are different,
shows that our previous analysis of the process by the fraction of
active collisions was erroneous [3,73]. The latter analysis had
sense only in the case of equal probabilities of each carbon atom

formation at any point of the growing surface, but it has now been experimentally proven that this does not occur.

The physicochemical features of PyC formation and the kinetic results considered in the present chapter permit certain conclusions to be made about the mechanism of this process. However, on the basis of these results it is not possible to construct a molecular mechanism of the process, that is, to point out the sequence of elementary reactions which altogether form the total process. This is due mainly to the absence of experimental methods to observe the process that takes place on the solid surface. We shall therefore restrict ourselves to the main conclusions and will not consider possible hypothetical molecular mechanisms, especially since this mechanism is apparently different for each hydrocarbon. These conclusions are, naturally, important for understanding the mechanism of the process and they must be taken into account in any future construction of its molecular mechanism.

The first-order kinetics of the total reaction of PyC formation and the first-order kinetics of the nucleation and growth of crystallites show that hydrocarbon adsorption is not a limiting stage of the process. Therefore, the use of the Langmuir equation for the description of the process has no physical sense. It should also be pointed out that Langmuir adsorption cannot explain hydrogen inhibition. First, hydrogen inhibition is quite different for various hydrocarbons. Second, a connection found for methane between hydrogen inhibition and the PyC structure shows that gas-phase hydrogen directly participates in the construction of a planar graphite layer and practically has no influence at all on the nucleation. Therefore, the use of the Langmuir equation for the description of hydrogen inhibition may be just formal.

As to the activation energy of PyC formation, it is the gross activation energy of a complicated process, in the course of which carbon-hydrogen bonds are broken and new bonds of a carbon-carbon graphite hexagonal lattice are formed. The activation energy of the total process of PyC growth equals the arithmetic average of the activation energies of nucleation and of crystallite growth.

But each of these activation energies is also the activation energy
of the total process. Therefore, the activation energies observed
experimentally cannot be directly connected with the energy of detachment of a hydrogen atom from a hydrocarbon molecule. In this
connection it may be noted that the greatest value of the activation
energy of nucleation observed for methane is 317 kJ mol^{-1}, which is
much smaller than the activation energy of the detachment of one
(427 kJ mol^{-1}) or two (377 kJ mol^{-1}) hydrogen atoms from a methane
molecule to form $CH_3^{\cdot}$ or $CH_2^{\cdot}$ radicals. For all the other hydrocarbons the activation energy of nucleation is much less than for
methane.

These considerations also prove that the formation of radicals
from hydrocarbons in the gaseous phase is not a limiting stage of
PyC formation and both processes, the nucleation and growth of crystallites, originate from stable molecules of the initial hydrocarbon.
However, this conclusion does not mean that the presence of hydrocarbon radicals which, together with stable molecules, strike the
growing surface of carbon has no effect on the total PyC growth rate.
On the contrary, the observed deviation from the first-order kinetics
of the reaction in a layer of a dispersed material at small partial
hydrocarbon pressures is explained by the presence of a superconcen-
tration of radicals in the gaseous phase which are formed from stable
molecules on the surface and whose destruction leads to the acceleration of nucleation and, consequently, to an acceleration of the total
PyC growth.

Naturally, these considerations refer only to those conditions
in which kinetic data were obtained. One may think that for higher
temperatures a contribution of radicals may be appreciable. This is
illustrated, for example, by the observed acceleration of the growth
of carbon black particles under the conditions of carbon black formation.

As has already been pointed out above, the molecular mechanism
of the nucleation and growth of crystallites is unknown. It may only
be stated that the mechanisms of these processes are different for
both the same hydrocarbon and for different hydrocarbons. It is

sufficient, for example, to note that for methane the activation energy of nucleation is 317 kJ mol^{-1} and that of the growth of crystallites is 227 kJ mol^{-1}, whereas for acetylene these values are 143 and 133 kJ mol^{-1}, respectively. Moreover, as has already been noted above, the different influence of hydrogen on the nucleation and growth of crystallites in the course of PyC formation from methane and the different effect of temperature on hydrogen inhibition for methane and acetylene provide a good illustration of the difference in the mechanisms of these processes.

In respect to the molecular mechanism of the nucleation and growth of crystallites the situation was approximately the same as when the chain theory of gaseous reactions had its genesis. It was quite clear that some special active particles provide this reaction, but the nature of these particles was unknown. One may state that crystallite nuclei are formed from stable hydrocarbon molecules which strike (from the gaseous phase) carbon atoms at the boundaries of crystallites and have an increased chemical activity. We do not know what these nuclei are nor how they are formed. It may only be said that these nuclei give rise to the beginning of the formation of a new graphite layer, again at the expense of hydrocarbon molecules of the gaseous phase.

Therefore, the growth of the layer occurs only on the periphery of growing crystallites. Molecules of the gaseous phase which strike carbon atoms of the formed graphite layer do not participate in the growth reaction.

With regard to the first-order kinetics of the nucleation reaction it should be thought that the nuclei formed in the course of PyC growth from methane differ from the nuclei which provide PyC growth from other hydrocarbons (e.g., from acetylene). However, it should be remembered that in spite of all this, these nuclei are quite interchangeable; that is the nuclei from acetylene provide growth of the graphite layer from methane not a bit worse than the nuclei formed from methane molecules. This is proved quantitatively by experiments measuring the rates of growth from binary mixtures.

XI. CONCLUSIONS

The values of the absolute constants of the rate of PyC formation obtained in the present work, and the kinetic model give, in principle, the possibility of calculating the rate of PyC formation for any process of hydrocarbon decomposition. They may serve as a basis for further quantitative investigations.

The author also hopes that the results obtained will promote further progress in the study of the formation processes of isotropic carbon and carbon black particles, and in prospect for the investigation of the molecular mechanism of all these complicated processes.

LIST OF SYMBOLS

C_V, C_U, C_W, C_L Preexponential multipliers of the Arrhenius equation for the PyC growth rate, the crystallite formation rate, the crystallite growth rate and the average crystallite size, respectively

d Distance between graphite layers, cm

E_L "Activation energy" of the average crystallite size, $J\ mol^{-1}$

E_V, E_U, E_W Activation energies of PyC growth, of the crystallite formation rate, and of the crystallite growth rate, respectively, $J\ mol^{-1}$

k $= V_2/V_1$

K Constant of the PyC formation rate, $g\ cm^{-2}\ s^{-1}\ Pa^{-1}$

K_U Constant of the crystallite formation rate, crystallite $cm^{-1}\ s^{-1}\ Pa^{-1}$

K_W Constant of the crystallite growth rate, $cm\ s^{-1}\ Pa^{-1}$

L_a Average size of a crystallite in the formation plane, cm

L_c Average size of a crystallite in the direction normal to L_a, cm

L_m Average size of a PyC crystallite from a binary mixture of hydrocarbons, cm

m	$= U_2/U_1$
n	$= W_2/W_1$
N	Number of crystallites, crystallite/cm^2
P	Perimeter of crystallites, cm/cm^2
U	Formation rate of crystallites (or nuclei), crystallite, $cm^{-1}\, s^{-1}$
V	PyC growth rate in the direction of a normal to the formation plane, cm/s
V_{US}	PyC growth rate from a binary mixture assuming the additivity of formation rates and growth of crystallites, cm/s
V_V	PyC growth rate from a binary mixture assuming the additivity of the film growth rates, cm/s
W	Growth rate of a crystallite in the formation plane, cm/s
α	Fraction of the second component in the binary mixture
β	V_{UW}/V_V
τ	Formation time of one monolayer, s

ACKNOWLEDGMENTS

The author is cordially thankful to Dr. A. E. Gorodetskii of the Institute of Physical Chemistry of the USSR Academy of Sciences, who wrote the sections on the methods of determining the PyC structure and made all the structural measurements. The author is also grateful to his co-workers from the laboratory for making all the kinetic measurements.

REFERENCES

1. H. B. Palmer and C. F. Cullis, in *Chemistry and Physics of Carbon* (P. L. Walker, ed.), Vol. 1, Marcel Dekker, New York, 1966.

2. J. C. Bokros, in *Chemistry and Physics of Carbon* (P. L. Walker, ed.), Vol. 5, Marcel Dekker, New York, 1969.

3. P. A. Tesner, *The Formation of Carbon from Hydrocarbons of a Gaseous Phase,* Khimiya, Moscow, 1972.

4. W. L. Kotlensky, in *Chemistry and Physics of Carbon* (P. L. Walker and P. A. Thrower, eds.), Vol. 14, Marcel Dekker, New York, 1973.

5. B. V. Derjaguin and D. V. Fedoseev, *Diamond and Graphite Growth from a Gaseous Phase,* Nauka, Moscow, 1977.

6. D. V. Fedoseev, S. P. Vnukov, and B. V. Derjaguin, *Carbon 17,* 453 (1979).

7. J. Lahaye and G. Prado, in *Chemistry and Physics of Carbon* (P. L. Walker and P. A. Thrower, eds.), Vol. 14, Marcel Dekker, New York, 1978.

8. G. Tamman, *Aggregatzustände,* Voss, Leipzig, 1922.

9. A. N. Kolmogorov, *Izv. Akad. Nauk SSSR, Ser. Mat.,* No. 3, 355 (1937).

10. L. M. Borodina and P. A. Tesner, *Khim. Tverd. Topl.,* No. 4, 157 (1983).

11. E. F. Arefieva, I. S. Rafalkes, and P. A. Tesner, *Khim. Tverd. Topl.,* No. 5, 113 (1977).

12. T. V. Tekunova and P. A. Tesner, *Khim. Tverd. Topl.,* No. 3, 147 (1974).

13. B. E. Warren and P. Bodenstein, *Acta Crystallogr. 18,* 282 (1965).

14. B. E. Warren and P. Bodenstein, *Acta Crystallogr. 20,* 602 (1966).

15. A. W. Kurdjumov and A. N. Pilankevich, *Zh. Strukt. Khim. 9* 859 (1968).

16. A. W. Kurdjumov and A. N. Pilankevich, *Phase Transformations in Carbon and Boron Nitride,* Naukova Dumka, Kiev, 1979.

17. P. A. Tesner, N. B. Golovina, A. E. Gorodetskii, and M. M. Poljakova, *Khim. Tverd. Topl.,* No. 1, 129 (1976).

18. P. A. Tesner, A. E. Gorodetskii, and T. V. Tekunova, *Khim. Tverd. Topl.,* No. 5, 118 (1975).

19. E. V. Denisevich and P. A. Tesner, *Khim. Tverd. Topl.,* No. 6, 126 (1977).

20. L. M. Borodina and P. A. Tesner, *Neftekhimiya 19,* 363 (1979).

21. L. M. Borodina and P. A. Tesner, *Neftekhimiya 20,* 388 (1980).

22. E. V. Denisevich and I. S. Rafalkes, *Khim. Tverd. Topl.,* No. 1, 119 (1978).

23. E. V. Denisevich and P. A. Tesner, *Neftekhimiya 20,* 390 (1980).

24. E. V. Denisevich, I. S. Rafalkes, and P. A. Tesner, *Zh. Fiz. Khim. 52,* 1567 (1978).

25. I. S. Rafalkes and P. A. Tesner, *Khim. Tverd. Topl.,* No. 2, 125 (1981).

26. L. M. Borodina and P. A. Tesner, *Khim. Tverd. Topl.,* No. 6, 140 (1978).

27. B. N. Altshuler, *Khim. Tverd. Topl.*, No. 4, 111 (1973).

28. S. S. Abajev and V. U. Shevchuk, *Gazov. Prom.*, No. 8, 33 (1965).

29. P. A. Tesner, M. M. Poljakova, and S. S. Mikheeva, *Dokl. Akad. Nauk SSSR 203*, 402 (1972).

30. E. V. Denisevich and P. A. Tesner, *Dokl. Akad. Nauk SSSR 212*, 660 (1973).

31. E. V. Denisevich and P. A. Tesner, *Khim. Tverd. Topl.*, No. 2, 142 (1974).

32. P. A. Tesner, A. E. Gorodetskii, and E. V. Denisevich, *Khim. Tverd. Topl.*, No. 3, 107 (1980).

33. P. A. Tesner, A. E. Gorodetskii, and E. F. Arefieva, *Kinet. Katal. 21*, 274 (1980).

34. P. A. Tesner, A. E. Gorodetskii, and L. M. Borodina, *Kinet. Katal. 23*, 990 (1982).

35. P. A. Tesner, A. E. Gorodetskii, L. M. Borodina, E. V. Denisevich, and A. G. Ljakhov, *Dokl. Akad. Nauk SSSR 235*, 410 (1977).

36. P. A. Tesner, A. E. Gorodetskii, and E. V. Denisevich, *Kinet. Katal. 22*, 774 (1981).

37. P. A. Tesner and I. S. Rafalkes, *Dokl. Akad. Nauk SSSR 87*, 821 (1952).

38. E. F. Arefieva and T. D. Snegireva, *Khim. Tverd. Topl.*, No. 3, 41 (1978).

39. E. F. Arefieva and T. D. Snegireva, *Zh. Fiz. Khimii 52*, 1839 (1978).

40. P. A. Tesner, A. E. Gorodetskii, T. D. Snegireva, and E. F. Arefieva, *Dokl. Akad. Nauk SSSR 239*, 901 (1978).

41. P. A. Tesner, T. V. Tekunova, and T. D. Snegireva, *Khim. Tverd. Topl.*, No. 4, 114 (1980).

42. P. S. Shantarovich and B. V. Pavlov, *Zh. Fiz. Khim. 34*, 960 (1960).

43. U. P. Jampolskii, *Dokl. Akad. Nauk SSSR 217*, 888 (1974).

44. V. G. Knorre, M. S. Kopilov, and P. A. Tesner, *Fiz. Goreniya Vzryva*, No. 5, 767 (1974).

45. M. M. Poljakova and P. A. Tesner, *Dokl. Akad. Nauk SSSR 93*, 885 (1953).

46. P. A. Tesner and I. S. Rafalkes, *Dokl. Akad. Nauk SSSR 80*, 401 (1951).

47. V. V. Keltzev and P. A. Tesner, *Carbon Black*, Gostoptekhizdat, Moscow, 1952.

48. J. B. Donnet and A. Voet, *Carbon Black*, Marcel Dekker, New York, 1976.

49. E. F. Arefieva and P. A. Tesner, *II Vses. Konf. Tekhnol. Goreniyu,* Chernogolovka, 1978.

50. P. A. Tesner, I. S. Rafalkes, and V. G. Knorre, *III Vses. Konf. Tekhnol. Goreniyu,* Vol. 1, Chernogolovka, 1981.

51. P. A. Tesner, V. G. Knorre, and T. D. Snegireva, *VI Vses. Symp. Goreniyu Vzr.,* Alma-Ata, *Kinet. Khim. Reak.,* Chernogolovka, 1980.

52. G. P. Churik, V. U. Shevchuk, and U. M. Romanjuk, *III Vses. Konf. Tekhnol. Goreniyu,* Chernogolovka, 1981.

53. H. B. Palmer, J. Lahaya, and K. C. Hou, *J. Phys. Chem. 72,* 348 (1968).

54. P. A. Tesner and B. N. Altshuler, *Dokl. Akad. Nauk SSSR 187,* 1100 (1969).

55. E. V. Denisevich, A. G. Ljakhov, and P. A. Tesner, *Zh. Fiz. Khim. 52,* 1566 (1978).

56. P. A. Tesner, A. E. Gorodetskii, E. V. Denisevich, and T. V. Tekunova, *Kinet. Katal. 23,* 1269 (1982).

57. R. J. Diefendorf, *J. Chem. Phys. 57,* 815 (1960).

58. P. A. Tesner, A. E. Gorodetskii, A. P. Zakharov, and M. M. Poljakova, *Dokl. Akad. Nauk SSSR 210,* 1379 (1973).

59. S. D. Robertson, *Carbon 8,* 365 (1970).

60. S. D. Robertson, *Carbon 10,* 221 (1972).

61. T. Baird, J. R. Fryer, and B. Grant, *Carbon 12,* 591 (1974).

62. F. J. Derbyshire, A. E. B. Presland, and D. L. Trimm, *Carbon 13,* 111 (1975).

63. T. Baird, *Carbon 15,* 379 (1977).

64. F. J. Derbyshire and D. L. Trimm, *Carbon 13,* 189 (1975).

65. G. Horz and K. Lindenmaier, *J. Less-Common Met. 35,* 85 (1974).

66. A. A. Kochergina, S. P. Vnukov, and D. V. Fedoseev, *Khim. Tverd. Topl.,* No. 6, 142 (1981).

67. S. L. Kharatjan, J. S. Sardarjan, A. A. Sarkisjan, and A. G. Merzhanov, *III Vses. Konf. Tekhnol. Goreniyu,* Vol. 1, Chernogolovka, 1981.

68. P. A. Tesner, A. E. Gorodetskii, and T. V. Tekunova, *Khim. Tverd. Topl.,* No. 3, 44 (1978).

69. T. V. Tekunova and P. A. Tesner, *Khim. Tverd. Topl.,* No. 5, 151 (1977).

70. P. A. Tesner, E. V. Denisevich, L. I. Efimov, and T. V. Tekunova, *Zh. Prikl. Khim. 52,* 955 (1979).

71. J. P. Hirth and G. M. Pound, *Condensation and Evaporation,* Pergamon Press, Oxford, 1963.

72. C. H. L. Goodman, ed., *Crystal Growth,* Vol. 1, Plenum Press, New York, 1974.

73. P. A. Tesner, *Eighth Symposium (International) on Combustion,* Combustion Institute, Pittsburgh, Pa., 1961.

3

Etch-decoration Electron Microscopy Studies of the Gas-Carbon Reactions

RALPH T. YANG

Department of Chemical Engineering
State University of New York at Buffalo
Amherst, New York

I. INTRODUCTION

The rates of gas-carbon reactions reported in the literature have
been measured mostly by thermogravimetric techniques and by evolved
gas analysis. The rate data, combined with the results achieved by
a wide variety of techniques, such as spectroscopy, isotope exchange,
temperature-programmed desorption, and thermoelectric power, have
provided the basis for our understanding of the mechanism and kinetics of the carbon gasification reactions. The rates thus measured
are the total rates on all surfaces of the carbon sample, which
include both active and nonactive sites, and are expressed on a g/g
basis.

Optical microscopy, which provides direct observation of the
reactions on the surface, has yielded many important results. The
most important one, perhaps, is the anisotropies of reactivity on
different crystallographic faces and the identification of the active
sites. An excellent review on the subject has been published by
Thomas [1]. Optical as well as scanning electron microscopy, however, do not yield kinetic data on the truly atomic scale (i.e.,
rates on an atom/atom active site basis). To achieve this resolution,
the etch-decoration electron microscopy technique was developed by
Hennig. The experimental details of this technique and Hennig's
pioneering work using the technique have been given in his seminal
paper [2].

It has been obvious since Hennig's work that a wealth of new
information could be obtained by the etch-decoration electron microscopy technique. This was indeed proved to be the case by the pub-
lications resulting from the use of this technique during the period
of 5 to 6 years following Hennig's untimely death in 1965. The most

significant work done during this period was that of Thomas, Feates, and co-workers on the $C + O_2$ reaction and that of Feates and others on the gasification by radiolytic gases. The experiment is difficult for beginners—in fact, temperamental. It was perhaps partly for this reason that no work using the technique was published during the decade following the paper by Evans et al. [3] until the research done in this laboratory.

The objective of this chapter is to provide a summary of the work done to date using the etch-decoration technique. This paper shall exclude the work done on catalyzed carbon gasification using the technique of controlled atmosphere electron microscopy invented by Baker and co-workers, and by scanning electron microscopy, as comprehensive reviews on these subjects are available [4,5].

II. EXPERIMENTAL TECHNIQUE

A. Underlying Principles

The two basic principles utilized by Hennig for the etch-decoration technique are the high anisotropy of reactivity of graphite on different crystallographic faces and the ability of the crystal edges to trap and nucleate gold. The anisotropy refers to the orders-of-magnitude higher reactivities on the edge atoms (e.g., on the zigzag or $\{10\bar{1}0\}$ and armchair or $\{11\bar{2}0\}$ faces) than on the basal plane (i.e., $\{0001\}$ faces). This anisotropy is caused by the free sp^2 electron on the edge atom. Due to the anisotropy, a single vacancy, which is surrounded by three edge atoms, can be expanded in reactive gases to form an etch pit which is one atomic layer deep. (Multiple vacancies, although fewer in number, can also be expanded into pits.) The edges of the pit can then be decorated by gold nuclei which are electron opaque and are visible in a transmission electron microscope. The principle and the main procedure of the experimental technique are illustrated in Fig. 1.

The single-crystal graphite that was found most suitable for etch decoration and handling (albeit not many samples have been tested) was the natural graphite from Ticonderoga in Essex County,

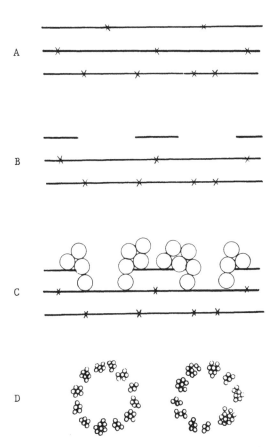

FIG. 1   (A) Schematic diagram of a graphite crystal normal to the
c-axis.   The crosses indicate the locations of lattice vacancies.
(B) The same crystal after a gasification reaction.   Monolayer pits
are formed.   (C) The pits are decorated by gold vapor deposited on
the surface.   The gold atoms (indicated by circles) nucleate on the
edges.   (D) Top view of the decorated sample in transmission electron microscope, where only gold nuclei are visible.   (From Ref. 3.)

New York [2].   The Ticonderoga graphite contains 10^{-8} to 10^{-7} vacancies or clusters per atom, although higher densities, as high as
2×10^{-6}, have been seen in our laboratory.   The origins of the
vacancies and other defects have been discussed in detail by Thrower
[6].   Upon etching or reaction, the vacancies, most of which are
probably single vacancies, lead to pits [7,8].   Thus far, there is

no direct proof that these etch pits are one atomic layer deep. Two
pieces of indirect, but convincing, evidence do exist. The first
was provided by Hennig [8], who created single vacancies by irradi-
ating the crystal with 0.8-MeV electrons, and the single vacancies
led to pits upon etching. The other evidence may be seen from the
merging pattern of the pits when the crystal is overetched, the
intersecting edges among pits always vanish, which is possible only
when all pits are of the same depth. It is highly improbable that
the depth is not monolayer.

B. Detailed Experimental Procedure

The experimental technique was, as mentioned, first established by
Hennig [2]. However, a detailed step-by-step description has not
been given. In the following a detailed procedure is described
which gives reproducible results and is not difficult to follow.
This procedure was established in this laboratory after a year of
painstaking trial and error, and was due largely to Wong [9].

Samples of Ticonderoga graphite may be obtained from Ward's
Science Establishment, Inc., Rochester, New York. The crystals of
graphite are liberated from their native marble matrix by dissolution
in HCl. The marble is broken into pieces and repeatedly leached
with reagent-grade HCl until the graphite flakes are freed. The
graphite flakes are further purified by alternate treatments of hot
HCl and HF; this procedure is repeated three times. Following the
acid treatment the samples are washed in distilled water for 90 min
and then dried at 100°C.

The graphite sample must be less than approximately 800 Å thick
to be observable in a transmission electron microscope. To produce
a sufficiently thin sample and to provide a clean basal surface to
carry out the gasification reactions, the sample must be successively
cleaved. To minimize the effect of impurities, the original faces of
the graphite crystals are not used in our experiments. A flake of
graphite is affixed to a clean glass slide with a small amount of
Duco cement and allowed to dry. On a separate slide a small drop of
Duco cement is spread to produce a very thin coating of cement. When

this cement film is tacky it is applied to the graphite sample, and gentle separation of the slides results in the formation of two fresh internal faces of graphite. Generally, these faces have a very smooth and shiny appearance. It should be noted at this point that the original faces of the crystal are both embedded in the cement on the glass slides; thus, repeating the procedure with these two samples will only produce samples from the interior of the crystal. The two original "halves" are cleaved to produce 10 to 20 samples using the technique described above and released from the glue by washing in reagent-grade acetone.

Promising samples (generally those with smooth surfaces) are affixed to slides with polyvinyl pyrrolidone (PVP), a water-soluble cement, with the desired face embedded in the cement. The PVP cement is produced by mixing 2 parts PVP powder (Kodak) with 1 part distilled water, by volume. The slide containing the cement is then applied to the graphite specimen. After the graphite is affixed to the cement a second slide is used to cover the graphite sample. The two slides are then clamped and dried in an oven at 60°C until the PVP is completely dried. Generally 2 h in the oven is sufficient to form a strong bond between the graphite and the glass slide.

The clamped specimens are removed from the oven and cooled to room temperature. Upon reaching room temperature the clamp and the cover slide can be removed and the specimen cleaved to its final thickness with Scotch tape. The cleaving procedure requires the Scotch tape to be carefully applied repeatedly, using tweezers, to remove layers or sections of layers of graphite. The sample is cleaved until a portion of the sample is translucent to light.

After final cleaving, the graphite samples are released by repeated washing with hot distilled water in a culture dish and removed with a piece of filter paper. These samples are then placed in a culture dish and dried at room temperature.

Graphite samples for noncatalytic reactions are transferred to the sample holder, which is then introduced into the reactor. For catalytic reactions, the samples are coated with catalyst particles, which may be introduced as slurry or solution. The procedure for coating catalysts has been given elsewhere [10].

To date, sapphire plate has been found to be the only acceptable sample holder. All other materials in contact with the graphite sample that we have tried, such as fused alumina, a porcelain combustion boat, a quartz plate, and so on, cause numerous channels on the graphite surface, apparently due to the impurities migrating onto the thin sample during the reaction. The sapphire plate is held in a porcelain combustion boat, which is in turn placed in a quartz reactor. The sapphire plate and the combustion boat are cleaned ultrasonically in acetone and subsequently distilled water each time before use. The gas purification procedures are different for O_2, H_2O, CO_2, and H_2. The procedures are given in detail elsewhere and will be referenced later for each individual gas-carbon reaction. Before the reaction, it has been found necessary and adequate to purge the sample in a purified argon or nitrogen flow for about 10 h at 650°C in the reactor. Inconsistent results and poor gold decoration will result without such a purge treatment, presumably because adsorbed species, if not desorbed, will retard the reaction and can also act as nucleation centers for gold. In most cases when purging was not done, no decorated pits, but random nuclei, are seen.

After the reaction, the sample is decorated with gold in a standard vacuum coater (Edwards 306 coater is used in our laboratory). The sample is held on a precleaned tungsten evaporation boat which is heated resistively to 250°C after the pressure in the coater has reached 10^{-5} torr. A section of gold wire is then evaporated on a tungsten filament 10 cm above the graphite sample. The amount of gold was precalculated to give one-monolayer coverage, assuming that the gold evaporates to yield a spherical distribution of vapor. At 10 cm approximately 4 mm of 0.008-in.-diameter gold wire is used. After coating, the sample is maintained at 250°C in vacuo for 20 min to facilitate the migration and subsequent nucleation of gold on the edges of the monolayer etch pits. The graphite is then removed from the coater and placed in a pair of copper folding grids for TEM examination.

    Selected area electron diffraction in the TEM is used, in situ,
to determine the orientation of the edges, steps, channels, and pits
(if they are not round). A molybdenum trioxide crystal is used to
calibrate the rotation between the diffraction pattern and the selected area image.

    The procedure described above has been used by four graduate
students in this laboratory to obtain consistent and reproducible
results. An etch-decorated TEM picture is shown in Fig. 2. This
picture is taken from a "clean" area; that is, no channels, edges,
and distortions of pits are seen in or near the area. Channels and
distorted pits which are produced by impurity particles do, however,

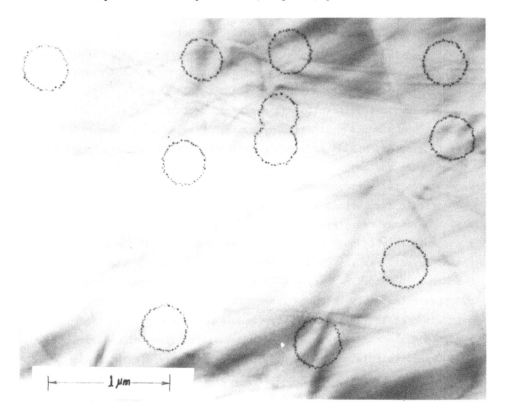

FIG. 2 Transmission electron micrograph of the basal plane surface
of graphite after oxidation for 25 min at 650°C with 0.2 atm O_2 followed by gold decoration.

exist in some areas of the graphite surface. Thus further sample
purification steps have been taken in this laboratory as well as by
all previous workers, as will be discussed shortly. The etch pits
in the clean areas are nonetheless not influenced and are reproducible, at least for the $C-O_2$ and $C-CO_2$ reactions which have been
studied thoroughly in this laboratory.

C. Effects of Additional Sample Purification

The sample purification procedure outlined above (i.e., repeated
leaching with hot HCl and HF solutions) does not give graphite samples of high purity. The impurities contained in these samples have
been measured as ash content and reported as 1 percent by Hennig [8]
and 0.2 percent by Thomas et al. [11]. Additional purification
treatments such as high temperature annealing and electron bombardment were undertaken by the subsequent workers to lower the ash content to the ppm level. In all of the work by Hennig on the $C-O_2$
reaction, a Cl_2/O_2 mixture was used as the etchant to oxidize the
graphite sample, following acid leaching without further sample
purification. The Cl_2 gas was used to remove the impurities, in
situ, on the surface during reaction, as recommended by P. L. Walker,
Jr. The Cl_2 gas was formed by saturating O_2 with CCl_4 and decomposing it. The gas composition in Hennig's work [2], according to our
calculation, was 0.26 atm Cl_2 and 0.74 atm O_2.

Hennig also measured the $C-O_2$ rates on samples further cleaned
by annealing at 3100°C and separately, by electron bombardment in
"high vacuum" at 2300°C, but noted that no difference was observed
in the loop expansion rates [8]. This result indicated that the impurities were generally not present on the freshly cleaved surfaces.
A further comparison can be made on the rates of the $C-O_2$ reaction
reported by several groups using the etch-decoration technique and
the same source of graphite. Turnover frequencies may be calculated
from this literature data and normalized to the conditions of 0.2 atm
O_2 and 650°C. The rates measured in this laboratory [12,13] using the
acid-leached samples and O_2 with no Cl_2 addition were 0.6 liter s^{-1}
(at 10 pits/μm^2) and 0.9 liter s^{-1} (at 1 pit/μm^2). (The turnover

rate for the C-O_2 reaction was found to depend on the pit density [12,13].) The rate determined by Hennig under these conditions, using Cl_2/O_2 and samples leached by acids with or without further purification, was 0.8 s^{-1} (pit density was not reported but was estimated at 5 pits/μm^2) [8]. The rate determined by Feates and Robinson with 0.2 atm (with no Cl_2 addition) at 600°C was 0.7 to 0.8 s^{-1} (pit density was 3 pits/μm^2 [14]). The sample cleaning procedure was not described in Ref. 14, but it was noted in related papers by Feates and his co-workers as electron bombardment at 2400°C in vacuo following washing in acids [15]. Using samples purified by the same method as that of Feates et al., Thomas and co-workers measured the rates of the C + O_2 reaction at 700 to 840°C in 1.33 kN/m^2 (10 torr) [3,16]. (The oxygen pressure was not clearly stated in Ref. 16, and was erroneously marked as 1.33 N/m^2 in Ref. 3. The value of 1.33 kN/m^2 is the only logical one as deduced from these references.) Using the square-root pressure dependence and the activation energy of 35 kcal/mole (as will be discussed later), their rate at 0.2 atm O_2 and 650°C was approximately 0.8 s^{-1}.

There is clearly a general agreement of the data among all four groups who have used the etch-decoration technique, regardless of whether additional sample purification was undertaken subsequent to the acid leaching procedure.

Additional sample purification may be desirable, nonetheless, because there are always areas on each sample which are contaminated and show channels and distorted pits. A simple and effective method has been found in this laboratory and is used for the C + H_2 reaction [17]. Prior to the argon purge at 650°C (for 10 h) and the subsequent reaction, the sample is treated with 0.26 atm Cl_2 in Ar at 650°C for 2 h, by saturating Ar with CCl_4 at 23°C and decomposing CCl_4 at 650°C. This procedure produces a very clean surface, but also yields sparsely distributed, hexagonal islands of graphite on the surface, presumably grown from the carbon generated by CCl_4 decomposition (Fig. 3).

D. Use of Scanning Electron Microscopy

The work done by Hennig and others preceded the commercialization of the scanning electron microscopy (SEM), and was thereby limited to

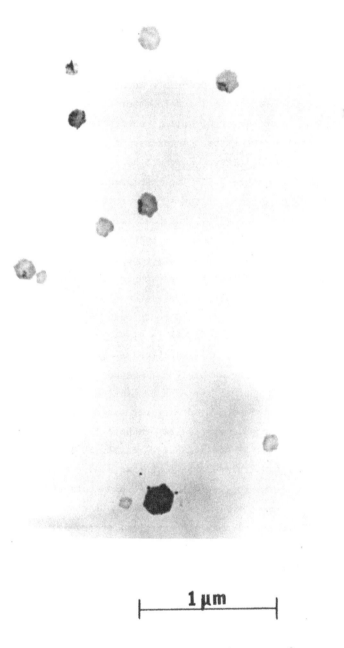

1 μm

FIG. 3 TEM of graphite after cleaning at 650°C for 2 h in CCl$_4$/He. This procedure eliminates channels formed by impurities during the subsequent reaction. Note the hexagonal carbon islands, with random orientation, formed from CCl$_4$ on the surface.

TEM. The use of SEM for etch-decoration studies has proved success-
ful in this laboratory [18]. Since cleavage of the graphite sample
to an ultrathin layer (for TEM) is not necessary for SEM studies,
the use of SEM makes the experimental procedure much simpler (and
easier) and also makes studies of noncleavable materials possible.

In the TEM studies, a monolayer of gold is enough to produce
resolved nuclei decorated on the steps. An excess of gold is necessary, however, to produce nuclei large enough to be resolved by SEM.
Consequently, a higher temperature (350°C) of the carbon surface was
required for mobility of gold atoms on the surface to form large
nuclei during the decoration step (250°C was used for TEM studies).
The optimum amount of gold for SEM studies of both carbon surfaces
was found to be seven monolayers. Exploratory experiments using
three monolayers did not produce nuclei large enough for SEM resolution; 10 monolayers produced excessive nucleation.

An SEM picture of the etched pits on Ticonderoga graphite decorated with seven monolayers of gold is shown in Fig. 4. The corresponding turnover frequency was 1.2 s^{-1} and agreed well with data on
thin samples examined by TEM. The sample for SEM study was cleaved
only once to provide a fresh surface for etch decoration.

The SEM technique has also been used with success on the oxidation of SP-1 graphite and the polycrystalline Graphoil (Union Carbide
Corporation), the latter being a noncleavable material. Besides the
turnover frequency on the basal plane, the technique also provides
additional microscopic information, such as the developments of
terraces, conical pits, caves, crevices, and so on, on the surface
[19]. The SEM technique certainly has the potential of revealing
new information on the mechanism and kinetics of gasification of
more complex carbon materials such as coal char.

III. THE C-CO_2 REACTION

This reaction is discussed first because it is the best understood
among the four gasification reactions and the results on this reaction will be used in the discussion of the other reactions.

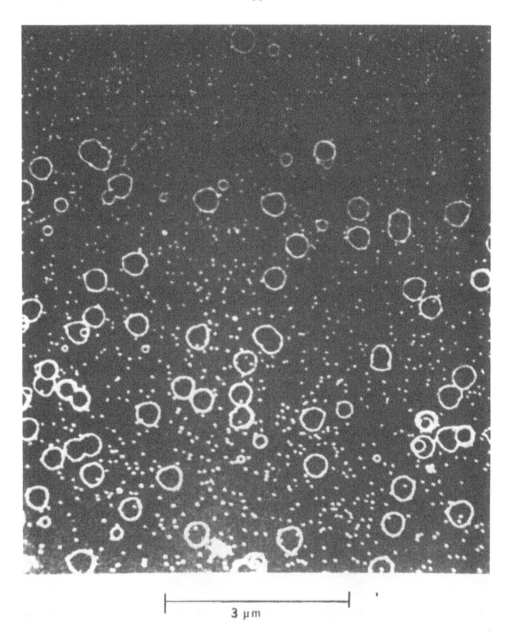

FIG. 4 Scanning electron micrograph of gold-decorated etch pits, created by oxidation in 0.2 atm O_2 at 650°C for 20 min on Ticonderoga graphite.

In studying the gasification reactions by H_2O, CO_2, or H_2, precautions must be taken to remove traces of O_2 from the "inert" carrier and the reactant gases. The ratio of rates as measured by TGA and evolved gas analysis (EGA) is $10^5 : 3 : 1 : 3 \times 10^{-3}$ for $O_2 : H_2O : CO_2 : H_2$, at 800°C and 0.1 atm, based on data using polycrystalline graphites and a low temperature char [20]. Thus the gasification rate by 1 ppm of O_2 is approximately the same as the rate by 0.1 atm of CO_2; and most of the commercial grades of the inert gases (e.g., N_2, Ar) and CO_2 contain more than 1 ppm of O_2. In this study an Oxygen-Free grade N_2 (<0.5 ppm O_2, Linde) was used as the carrier gas. The CO_2 was Aquarator Grade with a minimum purity of 99.99 percent, and the CO was Grade F with a minimum purity of 99.97 percent, both supplied by Linde Division of Union Carbide Company. The gas mixture was further purified by flowing through a preheater column packed with copper turnings maintained at 550°C to remove the residual O_2 in the gas stream. Oxidation of copper turnings by CO_2 is undetectable at this temperature, as concluded from our TGA measurements. In addition, a Drierite column was used to remove moisture in the gas mixture before the gas entered the reactor.

Two important gasification processes take place simultaneously on the basal plane of graphite: the removal of pit-edge atoms and the formation of new vacancies. The latter process creates new pits and is followed by the expansion of the new pits. The rates of both processes can be determined by the etch-decoration technique, and both are discussed below.

A. Elementary Rate Constants for Gasification
    of Monolayer Edge Carbon

The following Langmuir-Hinshelwood type of rate equation has been found to fit experimental data (all published data were obtained from TGA and EGA) for carbons ranging from dirty coal chars to high-purity single-crystal graphite [20-22]:

$$\text{Rate} = \frac{k_1 P_{CO_2}}{1 + k_2 P_{CO} + k_3 P_{CO_2}} \tag{1}$$

This can be derived from the most accepted form of the Langmuir-Hinshelwood mechanism:

$$C_f + CO_2 \; \underset{j_1}{\overset{i_1}{\rightleftharpoons}} \; C(O) + CO(g) \tag{2}$$

$$C(O) \; \xrightarrow{j_3} \; CO(g) \tag{3}$$

The rate constants in Eq. (1) are related to the elementary rate constants by $k_1 = i_1$, $k_2 = j_1/j_3$, and $k_3 = i_1/j_3$.

The turnover frequencies for the reaction have been measured by the etch-decoration technique at 600 to 900°C, $P_{CO_2} = 0$ to 0.8 atm and $P_{CO}/P_{CO_2} = 0.2$ to 0.8 [23]. The length of the reaction time to produce etch pits with sizes 0.1 to 1 μm ranged from 8 min to 8 h under these conditions. A typical transmission electron micrograph of a gold-decorated basal plane of graphite reacted with CO_2 is shown in Fig. 5, which gave a turnover frequency of 0.5 s^{-1}. The etch pits produced by CO_2 are circular, as in the C-O$_2$ reaction. But unlike the C-O$_2$ reaction, the pits are very uniform in size [24], which indicates that the surface diffusion mechanism is absent in the C-CO$_2$ reaction.

The elementary rate constants i_1, j_1, and j_3 are calculated from k_1, k_2, k_3 and are expressed as

$$\begin{aligned}
i_1 &= 9.07 \times 10^{10} \exp\left(-\frac{48,500}{RT}\right) \text{ s}^{-1} \text{ atm}^{-1} \\
&= 8.95 \times 10^{9} \exp\left(-\frac{48,500}{RT}\right) \text{ cm}^2 \text{ N}^{-1} \text{ s}^{-1} \\
j_1 &= 2.14 \times 10^{8} \exp\left(-\frac{33,200}{RT}\right) \text{ s}^{-1} \text{ atm}^{-1} \\
&= 2.11 \times 10^{7} \exp\left(-\frac{33,200}{RT}\right) \text{ cm}^2 \text{ N}^{-1} \text{ s}^{-1} \\
j_3 &= 4.85 \times 10^{7} \exp\left(-\frac{38,400}{RT}\right) \text{ s}^{-1}
\end{aligned} \tag{4}$$

These rate constants may be interpreted in terms of statistical or absolute rate theory as follows. In the forward reaction in reaction (2), we assume the CO_2 molecule is not localized in the adsorbed

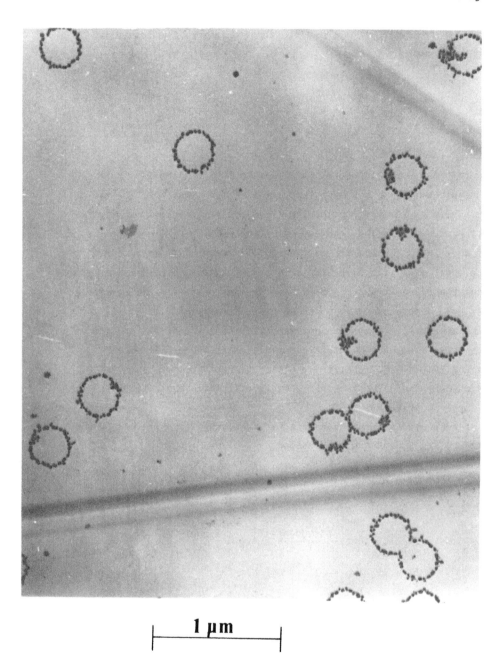

1 μm

FIG. 5 Transmission electron micrograph of gold-decorated monolayer etch pits formed by reaction of graphite with 0.8 atm CO_2 at 700°C for 55 min.

and activated state, and hence it possesses two degrees of translational freedom. Thus the frequency factor for i_1 is

$$i_{10} = \frac{5 \times 10^{13}}{RT} \frac{kT}{h} \frac{F^{\ddagger}}{F_{CO_2} f_s}$$

$$= \frac{4 \times 10^{13}}{RT} \frac{kT}{h} \frac{h}{(2\pi mkT)^{1/2}} \frac{b^{\ddagger}}{bCO_2} \qquad (5)$$

where F is the partition function for the species indicated by the
subscript, $\ddagger$ is the activated complex, f_s is the partition function
of active site, h is Planck's constant, k is Boltzmann's constant,
and R, T, and m have their usual meanings. In Eq. (5) $F^{\ddagger}$ is the
partition function of the activated CO_2 on the active site with $kT/h\nu$
factored out, or with one less degree of vibrational freedom. The
quantities denoted by b contain partition functions for rotational
and vibrational degrees of freedom ($b^{\ddagger}$ has no vibrational degrees
of freedom) for the activated and gas-phase CO_2. Equation (5) contains no concentration of active sites because i_{10} is expressed on
a per active site basis. The value of f_s is taken as unity since
the surface carbon atom has no translational or rotational degrees
of freedom and can undergo only very restricted vibration. If it is
further assumed that the adsorbed and activated CO_2 has two degrees
of rotational freedom, as does the gas-phase CO_2, we have

$$\frac{b^{\ddagger}}{b_{CO_2}} \cong 1 \qquad (6)$$

Here the vibrational factor for gas-phase CO_2 is near unity. Thus
the calculated value for the frequency factor is

$$i_{10} = 1.05 \times 10^7 \text{ cm}^2 \text{ N}^{-1} \text{ s}^{-1} \qquad (7)$$

which compares rather reasonably with the experimental value of
$8.95 \times 10^9 \text{ cm}^2 \text{ N}^{-1} \text{ s}^{-1}$.

Similar calculations may be made for the reverse reaction of
reaction (2). The frequency factor for the reverse reaction, j_{10},
can be calculated by assuming (1) the adsorbed and activated CO is

not localized, and (2) the chemisorbed O or C(O) is treated as a
localized active site. The calculated value is

$$j_{10} = 1.30 \times 10^7 \ cm^2 \ N^{-1} \ s^{-1}$$

which is in good agreement with the experimental value of 2.11×10^7
$cm^2 \ N^{-1} \ s^{-1}$.

The frequency factor for the net gasification reaction, C(O) →
CO, is given by statistical theory as

$$j_{30} = \frac{kT}{h} \tag{8}$$

assuming that both the activated complex and C(O) species are rigidly
held and the ratio of the partition functions is unity. Thus the
frequency factor is simply the vibrational frequency of the thoroughly
excited state of the C-C bond, which is about $2 \times 10^{13} \ s^{-1}$ in our
temperature range. This frequency is, however, for breaking of only
one C-C bond, whereas for the reaction to occur, two C-C bonds,
planar at 120° apart from the C-O bond, must be broken simultaneously.
The requirement for the correlated vibration of the two C-C bonds
should further lower the value for j_{30} from that for one bond. The
experimental value of $4.9 \times 10^7 \ s^{-1}$ is not unreasonable. In fact,
as seen in Table 1, all previous workers except Strange and Walker
reported a j_{30} value substantially lower than $10^{13} \ s^{-1}$, in agreement
with the discussion above.

B. Comparison of Rates on Multilayer
    and Single-Layer Edges

The active sites on carbon for gas-carbon reactions under normal
conditions are the atoms with a free sp^2 electron, which are the
surface atoms located on edges and defects. In all previous studies
the rates were measured by either TGA or evolved gas analysis, both
giving the total rates over all edges. The total rates were thus
contributed by predominantly multilayer edges. The rates measured
in this study, on the contrary, are rates on the monolayer edges or
steps.

For the $C-O_2$ reaction, the rate on the multilayer edge was reported to be about 100 times higher than that on the monolayer edge [3]. This enhancement phenomenon was termed the cooperative effect. More discussion on this effect will be given in the next section. It has been suspected, by comparing the ratio of rates for reactions with $O_2/H_2O/CO_2$ on multilayer and single-layer edges, that the cooperative effect does not exist for the reactions with H_2O and CO_2 [24]. With the data obtained in this study, it is now possible to make a direct comparison between the two rates for the CO_2 reaction.

Strange and Walker [22] reported TGA rates for a well-defined graphite sample (SP-1) in the temperature range 902 to 1007°C. The sample was of high purity and consisted of single-crystal disks with well-defined multilayer edges, which were the active sites responsible for their rates. A direct comparison can be made between their rates at 902°C and our rates at 900°C. The rate constants, in (this study)/(Strange and Walker) are approximately: for k_1, 99/0.05; k_2, 41/200; and k_3, 2/20, all in units containing atm and s, and the constants of Strange and Walker have been converted into C atom removed per C active site assuming that the SP-1 graphite was a perfect round disk of the dimensions given in the foregoing. The turnover frequency on the monolayer edge is thus about four orders of magnitude higher than the turnover frequency on multilayer edges as reported by Strange and Walker.

A direct comparison of our data on single-layer edges with other previous data as shown in Table 1 is not possible since the fraction of edge planes on the surface was not defined in their studies. However, it is apparent from Table 1 that our rates, mainly k_1, are higher than all reported rates by several orders of magnitude, because the fraction of edges in their samples was not likely to have been lower than that of SP-1 graphite, which is a rather large crystal.

From the comparison above it is clear that for the $C-CO_2$ reaction, the turnover frequency on monolayer edges is much higher than

TABLE 1 Comparison of Elementary Kinetic Parameters for the C-CO_2 Reaction

Carbon	i_{10} (s⁻¹ atm⁻¹)	E_1 (kcal/mole)	i_{20} (s⁻¹ atm⁻¹)	E_2 (kcal/mole)	i_{30} (s⁻¹)	E_3 (kcal/mole)	Reference
Pittsburgh bituminous coal coke	5.93×10^1	28.4	8.37×10^7	58.6	5.7×10^8	65.0	21
Spheron No. 6 Carbon Black	2.57×10^4	53.0	7.76×10^1	36.0	8.58×10^5	58.0	25
SP-1 graphite	1.2×10^{17}	99.0	3.2×10^{13}	74.0	4.35×10^{13}	87.0	22
Coconut charcoal	4.5×10^9	76.0	2.72×10^{11}	76.0	1.88×10^{10}	76.0	21
Activated carbon	—	—	—	—	1.43×10^7	59.0	26
Ceylon graphite	—	—	—	—	2.58×10^5	59.0	26
Coke	1.3×10^5	47.6	9.20×10^3	38.9	6.58×10^5	53.9	27
Anthracite	4.3×10^2	32.5	1.13×10^3	32.2	2.45×10^5	49.1	27
This work	9.07×10^{10}	48.5	2.14×10^8	33.2	4.85×10^7	38.4	

that on multilayer edges, contrary to the "cooperative effect" for the $C-O_2$ reaction. Although the cooperative effect for the $C-O_2$ reaction is not understood [3], it is possible to explain the reverse effect on the $C-CO_2$ reaction based on thermochemical considerations as discussed below.

Although the precise nature of the $C(O)$ species in the $C-CO_2$ reaction is not known, it is possible to estimate the bond strength between carbon and oxygen in this complex from the kinetic data. The heat of reaction of the oxygen exchange reaction [i.e., C_f + $CO_2 \rightleftharpoons C(O) + CO$] is $E_1 - E_2$, which is equal to $48.5 - 33.2 = 15.3$ kcal mol^{-1} endothermic, from the data given previously. The bond energy between C and O in $C(O)$ can be calculated since the bond to be formed (in CO) and the bond to be broken (in CO_2) are both known. However, the formation of $C(O)$ on the edge carbon involves some localization of a π electron from the graphite, which results in a loss of C-C bond energy within the graphite, as shown by Walker and co-workers [28]. The bond energy in $C(O)$ can thus be estimated by considering these three energies. Based on a heat of reaction for the oxygen exchange reaction of 25 kcal mol^{-1} endothermic as calculated from their TGA rate data, Strange and Walker [22] estimated that the $C(O)$ bond has two-thirds double-bond character. Their value is for the oxygen chemisorbed on the multilayer edge. Our value for the heat of reaction of 15.3 kcal/mole indicates that the $C(O)$ bond on the monolayer edge is about 9.7 kcal/mole stronger than that on the multilayer edge. The stronger $C(O)$ bond is apparently the reason for (1) the higher rate of formation of $C(O)$, and (2) the easier breakage, and hence the higher rate, of $C(O)$ from the graphite (with two C-C bonds to its neighbors).

The substantially higher rates of the CO_2 reaction on the monolayer edge as compared with those on the multilayer edge cannot be attributed to impurities (or catalysts) because most of the carbon samples listed in Table 1 contained considerably more impurities than the graphite sample used in this study, and the effect is reversed for the $C-O_2$ reaction.

The monolayer edge and multilayer edge represent two extreme types of active sites for the gas-carbon reactions. The "cooperative effect" in the $C-O_2$ reaction is likely to be a gradual one depending on the number of layers forming the edge, as seen in our laboratory [29]. The dependence of reactivity on the number of layers forms the basis for the familiar compensation effect; that is, the frequency factor and the activation energy for a given surface reaction usually increase or decrease together. The compensation effect for the $C-CO_2$ reaction is evident in Table 1. Our interpretation of the compensation effect is in accord with Sosnovsky's theory on catalytic reactions [30].

C. Rate and Mechanism of Vacancy
    Formation on the Basal Plane

In gas-carbon reactions two concurrent surface processes take place: (1) removal of edge carbon atoms on both multilayer edge and monolayer edge, and (2) abstraction of carbon atoms within the basal plane, which is followed by the formation and expansion of new pits. Vacancy formation refers to the second process. It is known that the second process is much slower than the first one and contributes relatively little to the overall rate. It is important, however, to understand this process because it becomes important at higher temperatures. Hennig [8] and Evans and Thomas [16] have measured the rate of vacancy formation for the $C-O_2$ reaction.

The rate of vacancy formation for the $C-CO_2$ reaction is determined by the number of pits which are smaller than the group of the largest pits. The largest pits are uniform in size and originate from the inherent vacancies. The smaller pits with varying sizes, are from the vacancies formed during the reaction. A typical micrograph of pits formed from new vacancies is shown in Fig. 6. It is also possible to determine from such micrographs the rate of vacancy formation as a function of reaction time.

The rate of vacancy formation was measured for temperatures of 600 to 900°C. The abstraction of basal plane atoms becomes vigorous only at temperatures above about 700°C. Figures 7 and 8 present, respectively, the rates of vacancy formation at 900°C for $CO_2 = 0.4$

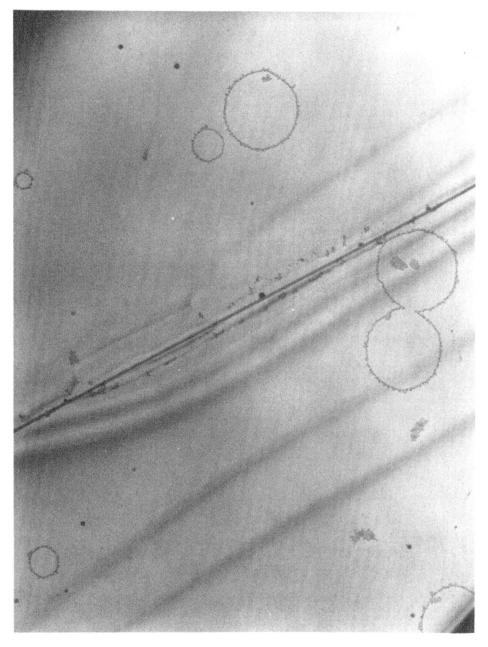

FIG. 6 Transmission electron micrograph of gold-decorated etch pits on graphite etched by CO_2 showing smaller pits originated from newly formed vacancies.

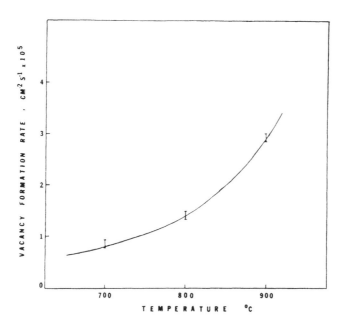

FIG. 7 Rate of vacancy formation in graphite lattice at 900°C in
0.4 atm pressure of CO_2 with no CO addition.

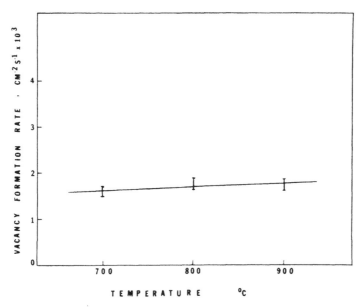

FIG. 8. Rate of vacancy formation in graphite lattice at 900°C in
0.4 atm P_{CO_2} and P_{CO}/P_{CO_2} = 0.4.

atm (without CO) and for 0.4 atm CO_2 with P_{CO}/P_{CO_2} = 0.4 (at a total pressure of 1 atm). With even a very small amount of CO addition, the rate of vacancy formation is drastically lowered. In such cases it becomes very difficult to count the rate because a very large surface is needed.

The foregoing data can be accounted for from kinetic considerations by assuming that the O atoms dissociated from CO_2 are responsible for the abstraction. The equilibrium concentration of O atoms for the thermal dissociation

$$CO_2 \rightleftarrows CO + O \tag{9}$$

can be calculated [23] and the frequency of collision of the O atoms on the basal plane carbon can also be calculated [23]. The probability of abstraction of a carbon atom upon each collision by O atom is

$$\varepsilon = \frac{\text{rate of vacancy formation}}{\text{collision frequency}}$$

$$= 1.13 \times 10^{-10} \ (700°C); \quad 1.81 \times 10^{-10} \ (800°C);$$

$$3.82 \times 10^{-10} \ (900°C) \tag{10}$$

The values of ε calculated for the C-O_2 (based on the data from Refs. 8 and 16) and the C-H_2O reactions [31] are very close to the values above for the C-CO_2 reaction. This is a strong evidence that the O atoms, regardless of their origin, are responsible for vacancy formation. Furthermore, with the addition of CO, the concentration of O atoms is drastically reduced in reaction (9), which results in the observed reduction of the vacancy formation rate.

IV. THE C-O_2 REACTION

As for the C-CO_2 reaction, the results to be discussed in this section will be limited to those obtained directly from and related to the etch-decoration experiments.

A. Rate Equation

As is clear from the comparison made in Sec. II, the rates measured by four different groups for the C-O_2 reaction using etch decoration

are essentially the same, regardless of the sample cleaning procedures used. The rates can be summarized by the following equation:

$$\text{Rate} = k_0 P_{O_2}^{1/2} e^{-E/RT} \quad \text{s}^{-1} \tag{11}$$

where $k_0 = 3.2 \times 10^8$ s^{-1} atm$^{-1/2}$, and E = 35 kcal/mole. It should
be emphasized that this equation is based on data in the temperature
range 600 to 800°C and O_2 pressures below 1 atm.

Equation (11) gives the turnover frequency for the gasification
rate on monolayer step atoms on the basal plane with a pit density
in the range of approximately 1 to 5 pits/μm^2, which is the normal
vacancy density in Ticonderoga graphite.

It is important at this point to compare the monolayer-edge
data, via Eq. (11), with the multilayer-edge data, which are obtained
by TGA or EGA. It is also important to discuss the one-half-order
pressure dependence.

B. Comparison of Rates on Multilayer and Single-Layer Edges

The comparison of rates on multilayer and single-layer edges for the
$C-O_2$ reaction is not as clear and definitive as that for the $C-CO_2$
reaction. It would be convenient for us first to summarize briefly
the comparisons for the $C-O_2$ reaction.

The turnover frequency on the multilayer edges is higher than
that on the single-layer edges, by approximately one order of magnitude, in the temperature range 600 to 800°C. This summary is based
on comparisons using available data in the literature and a direct
comparison with the TGA data obtained in this laboratory, which will
be shown in this report.

The first comparison was made by Thomas and co-workers [3], who
concluded that at 840°C, the recession rate of the multilayer edge
was nearly 100 times higher than that on a single-layer edge as measured by etch decoration. The multilayer edge recession rate was
evaluated from a series of optical micrographs, and the reaction
conditions were at a low O_2 pressure (1.33 N/m^2) and at 812 to 872°C.
The activation energy for the multilayer recession in the temperature

range above was 64 ± 4 kcal/mole. This result cannot be extrapolated
to temperatures lower than 800°C.

It is possible to calculate the turnover frequency on multilayer
edges from the EGA data obtained by Walker and co-workers [32]. In
this work, the active sites on a Graphon sample were titrated by chemisorption of O_2 (at an initial pressure of 500 μ) at 300°C. The
amount of chemisorbed O_2 was measured by desorption at 950°C. From
their rate data, the turnover frequency at 625°C and 0.2 atm O_2 was
approximately 30 s^{-1}. This turnover frequency is approximately 60
times higher than that on the monolayer edge measured by etch decoration.

A good estimate of the turnover frequency on multilayer edges
can be made by using rate data on SP-1 graphite. This sample is a
single crystal with rather uniform dimensions, as described above,
and it has been used quite fruitfully by Walker and co-workers. To
obtain first-hand data we have measured the TGA rates of O_2 oxidation
using SP-1 graphite. These data were then compared directly with our
etch-decoration data.

The TGA data were gathered using a Mettler TA-2000C thermo-
analyzer. To eliminate the catalytic effect by the sample holder,
a thin layer (about a 10-mg sample) of SP-1 graphite was placed on
a sapphire plate which was mounted on an alumina sample holder.
Rates were measured at 10 to 15 percent burnoff, 650 to 850°C, 0.08
to 0.5 atm O_2 [33]. Assuming that the sample was composed of uniform
circular disks with initial thickness 0.4 μm and diameter 30 μm, and
that all active sites were on the peripheral edges, the TGA data were
converted into turnover frequencies as shown in Table 2. The turnover frequencies obtained here were high estimates because the SP-1
disks were not perfectly circular and there were many active sites
on the "basal" planes in the forms of terraces and defects. Nevertheless, the turnover frequencies on the multilayer edges were higher
than those on the monolayer edges, probably by a factor of less than 4.

The enhanced reactivity toward O_2 of the multilayer edges, or the
"cooperative effect," has not been explained. It is possible to offer
the following discussion based on the results on the C + CO_2 reaction.

TABLE 2 Comparison of Turnover Frequencies (TOF) on Multilayer
and Single-Layer Edges at O_2 = 0.36 atm

	Temperature (°C)				
	650	700	750	800	850
TOF, multi. (s^{-1})	4.8	12.5	43.5	140	400
TOF, mono. (s^{-1})	1.1	3.0	7.1	15.8	—
Multi./mono.	4.4	4.2	6.1	8.9	—

Although the mechanism of the $C + O_2$ reaction is not understood,
and is more complex than the $C + CO_2$ reaction, the following two
general steps are involved:

$$C + O_2 \rightleftarrows C(O) \tag{12}$$

$$C(O) \rightarrow CO \tag{13}$$

The actual mechanism is far more complex than the one above. For
example, there are many possible species of the chemisorbed oxide
(e.g., C_2O, CO_2, etc.) on various types of edge sites, and the net
carbon gasification step may also involve an Eley-Rideal type of
mechanism [i.e., $C(O) + O_2 \rightarrow CO$ or CO_2]. The mechanism above simply
indicates that the reaction involves two major steps: chemisorption
and rupture of carbon-carbon bonds.

Based on the thermochemical data for the $C + CO_2$ reaction, it
has been seen that the $C(O)$ bond on the monolayer edge is about 8.7
kcal/mole stronger than that on the multilayer edge. A similar situation may be expected in the $C + O_2$ reaction. Thus the higher rate
on the multilayer edge (than that on a monolayer edge) for the $C + O_2$
reaction is due to a higher rate in the chemisorption step, not in
the C-C bond breaking step. In other words, the rate of chemisorption of O_2 is higher on the multilayer edge than on the monolayer
edge, which results in a higher overall rate on the multilayer edge.

C. Comparison of Reaction Orders and Activation
    Energies for Multilayer and Single-Layer Edges

The pressure dependence and the activation energy for the $C + O_2$
reaction both vary widely in the literature data. The pressure

dependence falls mostly in the range 0.5 to 1, and the activation energy, 40 to 60 kcal. These data are for multilayer edges. Our TGA data using SP-1 graphite are shown in Table 3. These kinetic parameters vary with temperature. It can also be seen in Table 3 that the parameters are rather constant for the monolayer edges. The varying reaction order and activation energy for the overall $C + O_2$ can be expected, as shown by the following example.

The mechanism

$$C + \frac{1}{2} O_2 \rightleftarrows C(O) \tag{14}$$

$$C(O) + \frac{1}{2} O_2 \rightarrow CO + C(O) \tag{15}$$

$$C(O) \rightarrow CO \tag{16}$$

$$C(O) + O_2 \rightarrow CO_2 + C(O) \tag{17}$$

would result in a rather complex rate equation:

$$\text{Rate} = \frac{k_1 P^{1/2} + k_2 P}{1 + k_3 P^{1/2}} \tag{18}$$

not even considering the different structures and carbon sites of the chemisorbed oxygen. A multitude of mechanisms similar to the one above can be written, each leading to a different rate equation.

TABLE 3 Comparison of O_2 Pressure Dependence (Order) and Activation Energy (E) Between TGA Data (on SP-1 Graphite) and Etch-decoration (ED) Data

	Temperature (°C)				
	650	700	750	800	850
Order (TGA)	0.51	0.53	0.82	0.91	1.02
Order (ED)	0.5	0.5	0.5	0.5	0.5
E (TGA) (kcal)	34			34	
E(ED) (kcal)	35	35	35	35	—

Source: Ref. 33.

The data in Table 3 indicate that the mechanism on the multi-layer edges is changing with temperature, whereas it appears to remain unchanged for the monolayer edge in the temperature range studied. This comparison may indicate that a rather simple mechanism is in force for the $C + O_2$ reaction on the single-layer edge, such as

$$C_f + \frac{1}{2} O_2 \overset{i_1}{\underset{j_1}{\rightleftharpoons}} C(O) \tag{19}$$

$$C(O) \overset{j_3}{\rightarrow} CO \tag{20}$$

which gives

$$\text{Rate} = kP^{1/2} \tag{21}$$

where $k = j_3 i_1/(j_1 + j_3)$ and $j_1 + j_3 \gg i_1 P^{1/2}$. The simple form in Eq. (21) agrees with the experimental rate equation [Eq. (11)], although the actual mechanism may be more complex.

D. Surface Diffusion in the $C-O_2$
    Reaction Mechanism

Surface diffusion refers to the well-known phenomenon that the ad-sorbed species can be mobile on the surface. The role of surface diffusion in surface reaction mechanisms is, however, not clear, since no unequivocal evidence on such a role has been shown.

For the $C-O_2$ reaction at 650°C and 0.2 atm O_2, two independent sets of experiments have demonstrated the importance of surface dif-fusion in the overall reaction rate. The experimental evidence is as follows [12,13]: (1) the dependence of turnover frequency on the density of active sites, and, (2) the continued carbon gasification after O_2 is cut off from the gas flow over the surface.

The site-density dependence of the turnover frequency is shown in Fig. 9. As mentioned in the preceding section, the rate of the $C-CO_2$ reaction does not depend on the site density. The continued carbon removal after the elimination of O_2 in the gas phase is shown in Fig. 10. Note that the continued gasification is dependent on the density of the etch pits on the basal plane, and it is more

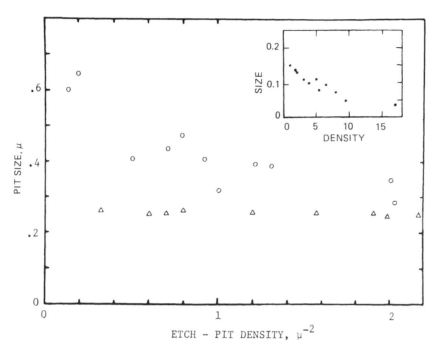

FIG. 9 Dependence of the pit size, which is proportional to the turnover frequency of carbon gasification, on the density of pits, for the C-H$_2$O reaction (23 torr H$_2$O, 600°C, 14 h) (o), the C-CO$_2$ reaction (1 atm CO$_2$, 600°C, 2 h) ($\triangle$), and the C-O$_2$ reaction (0.2 atm O$_2$, 650°C, 20 min) (upper right corner). (From Ref. 24.)

vigorous for surfaces with a sparse pit density. The results in Fig. 10 show that there is oxygen stored on the surface which continues the gasification of the pit-edge carbon atoms after the exhaustion of gas-phase O$_2$, through a surface diffusion mechanism. The results in Figs. 9 and 10 may be used to calculate the amount of stored oxygen on the surface, the surface diffusion coefficient, and the rate constant of the active site reacting with the oxygen arriving on the surface [13]. The results are shown in Table 4.

These results showed unequivocally that surface diffusion of oxygen is operative in the C-O$_2$ reaction. It is not clear at this point, however, whether the basal-plane carbon atoms are the sites for chemisorbing oxygen. It may be possible that the surface oxygen

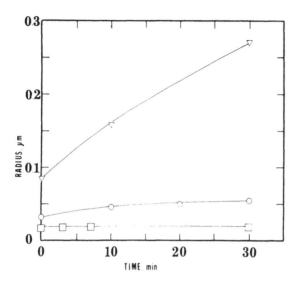

FIG. 10 Ring radii versus time of flushing in argon after a 10-min
reaction in 0.2 atm O_2 (0.8 atm Ar) at 650°C. Ring density (μm^{-2}) =
1 ($\triangledown$), 10 ($\circ$), and >20 ($\square$). Time zero corresponds to the time when
O_2 is cut off. (From Ref. 13.)

is adsorbed by impurities, defects, and ledges, all existing on the
surface. In that case, these chemisorption centers must be present
in an unreasonably high concentration to account for the large amount
of carbon removal.

TABLE 4 Surface Rate Constants, Surface Diffusion Coefficients, and
Amounts of Chemisorption (After 10 min Reaction) for Oxygen on (0001)
Plane of Graphite, at 0.2 atm O_2 and 650°C

Ring density (μm^{-2})	k (s^{-1})	D ($cm^2\ s^{-1}$)	Chemisorption (O/C)
1	4.14	5×10^{-12}	0.34
10	4.49	2×10^{-12}	0.06
>20	—	—	Nil

Source: Ref. 13.

E. Catalyzed Reaction on Monolayer Edges

The behavior of catalyst particles during the catalyzed carbon gasification is rather unique in that the particles form channels or
localized deep pits on the carbon substrate. This phenomenon has
been revealed and studied on the basal plane of graphite single
crystals using etch-decoration electron microscopy [34], controlled
atmosphere hot-stage optical microscopy [1,33], and electron microscopy [4,36]. The most powerful tool for this research is the controlled atmosphere electron microscopy [4,37]. With a few exceptions,
all channeling studies have been done with the carbon-oxygen reaction,
in the temperature range 500 to 800°C. Over 30 metals and metal
oxides, mostly transition and noble metals, have been studied [1,4,
34-37]. Most of the catalyst particles form channels, while a few
form pits or both channels and pits. There are, however, several
catalysts which do not form channels or pits, yet they do catalyze
the oxidation reaction. The etch-decoration technique provides a
unique tool for studying the catalytic actions of these nonchanneling,
nonpitting catalysts. Among this type of catalyst are Ta_2O_5 [1],
Cr_2O_3 [35,38], and WC [39]. A more complete study of these catalysts
was conducted in this laboratory on seven transition metal carbides
and oxides for the $C-O_2$ reaction at 680°C with 0.2 atm O_2 [40]. Only
MoO_3 was found to be a channeling catalyst, due to its low melting
temperature (837°C, or Tamman temperature = 282°C). The carbides and
oxides all proved to be effective catalysts, but showed distinctively
different mechanisms between the two groups.

The carbide group investigated included WC, TaC, and Mo_2C. The
turnover frequency was increased substantially over that of the uncatalyzed reaction (Table 5). More important, the etch pits were all
hexagonal in shape, and the sides of the pits were composed of zigzag
faces, $\{10\bar{1}0\}$, as identified by selected area diffraction using TEM.
Another important feature of the transition metal carbide catalyzed
reaction was that no channeling or pitting was observed. The catalyst
particles were at distances away from the etch pits. The catalysis
was apparently due to a remote-action mechanism.

TABLE 5 Overall Turnover Frequencies for Carbon Oxidation
(Over 7 to 10 min Reaction Time) in 0.2 atm O_2 at 680°C
Catalyzed by Transition Metal Carbides and Oxides

Catalyst	Overall turnover frequency (s^{-1})	Tammann temperature (°C)
WC	6.2	1325
TaC	12.5	1887
Mo_2C	12.5	1266
WO_3	15.6	635
Cr_2O_3	4.7	1135
Ta_2O_5	3.1	804
MoO_3	Formed channels	282
None	0.7 - 1.2	—

Source: Ref. 40.

The pits produced on the basal plane by the WC-catalyzed carbon-
oxygen reaction are shown in Fig. 11. The overall turnover frequency
calculated from this figure was 6.2 s^{-1}. An uncatalyzed sample oxidized under the same conditions has a turnover frequency in the range
of 0.7 to 1.2 s^{-1}. The WC particles retained the same sizes and are
not shown in the figure. The other carbides tested (TaC and Mo_2C)
behaved in a similar way to WC [40], and the catalyzed turnover frequencies are shown in Table 5.

Our experiments with graphite gasification by atomic oxygen have
shown that the resultant monolayer etch pits were hexagonal, and the
orientation of the hexagonal pits was that with the zigzag faces [41].
An interpretation has been given on the different conformations of the
etch pits formed by O_2 and by O atoms [41]. Figure 12 illustrates the
early development of the etch pit starting from a single vacancy. It
is seen that after the first three layers of active sites are removed,
the pit is already bonded by the surface atoms which form a hexagon,
and the sides of which are the zigzag, or $\{10\overline{1}0\}$ faces. Further removal of the active sites, layer by layer, will simply expand the
hexagonal pit. To form a circular pit, on the other hand, will require

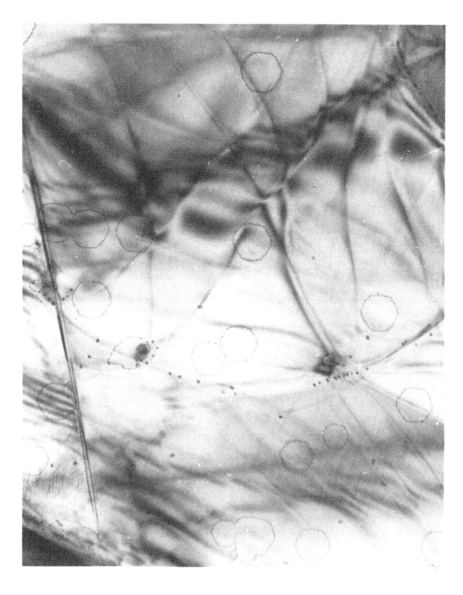

—3 μm—

FIG. 11 Hexagonal etch pits with the zigzag {10$\bar{1}$0} faces produced on the basal plane of graphite by oxidation at 680°C in 0.2 atm O_2 for 10 min with WC catalyst. Two-micron-size WC particles are not shown in the figure. (From Ref. 40.)

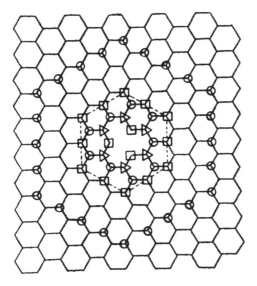

FIG. 12 Conformation of etch pit from a single vacancy in the early
development. The symbols denote surface atoms at each step during
pit expansion. (From Ref. 24.)

additional removal of atoms near the middle of each side, or making

the hexagon rounded. Once rounded, continued removal will lead to

a larger circular pit. Thus the conformation of etch pits is deter-

mined in the early stage of pit expansion. In the early stage, after

a few layers have been removed, the dimension of the pit (a few ang-

stroms) is relatively small compared to the collision diameter of the

reactant gas molecules (about 4 Å for O_2), and therefore the proba-

bility of removal of the corner atoms in the hexagonal pit is hindered

due to steric reasons.

The hexagonal shape of the etch pits, with the zigzag orientation,

for oxidation catalyzed by transition metal carbides is indicative of

attack by atomic oxygen. The transition metal carbides have electronic

structures similar to that of platinum and possess catalytic activities

for dissociating O-O and H-H bonds [42]. Therefore, it is likely that

the transition metal carbides act as dissociation centers for molecular

oxygen generating atomic oxygen or surface oxide species which subse-

quently diffuse across the basal plane of graphite and react at edges

of the monolayer pits.

The behavior of the oxide catalysts (WO_3, Cr_2O_3, and Ta_2O_5) is very different from that of the carbide group [40]. The oxide group all produced circular, but deformed etch pits. The estimated turnover frequencies of the oxide-catalyzed reaction are shown in Table 5. A sample whose reaction was catalyzed by Ta_2O_5 is shown in Fig. 13. The pits from reactions catalyzed by WO_3 and Cr_2O_3 are less severely deformed than for the Ta_2O_5-catalyzed case [40]. Another common feature of the oxide-catalyzed reaction was that the oxide particles, with an original size of about 0.2 μm, all disappeared during the oxidation step, and they dispersed into small particles with sizes of the order of nanometers [43].

The experimental results described above and the fact that the reaction temperature (680°C) was near or above the Tammann temperatures of the catalysts suggested a mechanism in which molecular species or small clusters were emitted from the particles and were trapped on the edges of the pits, and subsequently catalyzed the reaction by direct contacts. The breakup and spreading of the particles on the carbon surface might also be caused by wetting and spreading. The uneven distribution of the trapped particles on the edges of the pits would cause the deformed pits. The mechanism of the catalyzed reaction by the trapped particles or clusters may be either one of the mechanisms proposed for the catalyzed $C-O_2$ reaction by catalysts in direct contact with the carbon surface [5].

A more recent study of the oxide-catalyzed reaction showed that the turnover frequency on the edge carbon atoms was independent of the amount of the catalyst loading on the surface, and the activation energies were 27.5 kcal/mole for the WO_3 catalyzed reaction, and 29.5 kcal/mole for the Cr_2O_3 catalyzed reaction [43], as compared with 35 kcal/mole for the uncatalyzed reaction.

V.   THE $C-H_2O$ REACTION

A.   Published Rate Equations and Mechanisms

Despite the fact that the $C-H_2O$ reaction is a more complex one than the $C-CO_2$ reaction, it has been less intensively studied. Early

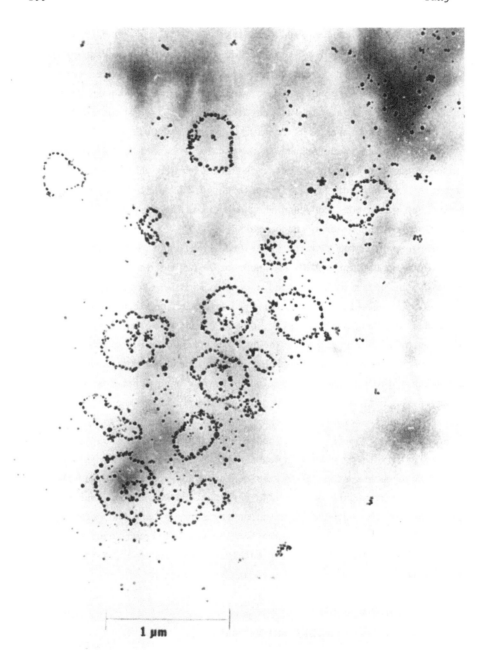

1 μm

FIG. 13 Etch loops produced by oxidation at 680°C in 0.2 atm O_2 for
10 min with Ta_2O_5 catalyst initially deposited on surface. Note the
formation of the severely deformed pits. (From Ref. 40.)

experimental data on the $C-H_2O$ reaction fit the following rate equation [20]:

$$\text{Rate} = \frac{k_1 P_{H_2O}}{1 + k_2 P_{H_2} + k_3 P_{H_2O}} \qquad (22)$$

A number of mechanisms can be written all leading to Eq. (22). One of these mechanisms is the one similar to the mechanism for the $C-CO_2$ reaction [20]:

Mechanism B:

$$C + H_2O \underset{j_1}{\overset{i_1}{\rightleftarrows}} C(O) + H_2 \qquad (23)$$

$$C(O) \overset{j_3}{\rightarrow} CO \qquad (24)$$

According to mechanism B, the retardation by H_2 is caused by the chemisorption and reaction of H_2 on the chemisorbed oxygen. It is entirely possible that H_2 may also chemisorb on the carbon active site free of oxygen, which will then lead to a different rate equation [44], and will also have a possibly strong retardation effect on the $C-CO_2$ reaction because the $C-CO_2$ reaction is slower than the $C-H_2O$ reaction [45].

A study of the $C-H_2O$ reaction using a tenfold wider range of partial pressures of added hydrogen than the earlier studies gave the following rate equation [44]:

$$\text{Rate} = \frac{k_1 P_{H_2O}}{1 + k_2 P_{H_2}^{1/2} + k_3 P_{H_2O}} \qquad (25)$$

The temperature range for this study was 790 to 850°C. A polycrystalline graphite (TSX from National Carbon Company) was used and the rate was measured gravimetrically. The partial pressures of H_2O and H_2 were both kept below 10 torr. The corresponding mechanism for Eq. (25) is [44]:

Mechanism C:

$$C + H_2O \xrightarrow{i_1} C(O) + H_2 \tag{26}$$

$$C + \frac{1}{2} H_2 \underset{j_2}{\overset{i_2}{\rightleftarrows}} C(H) \tag{27}$$

$$C(O) \xrightarrow{j_3} CO \tag{28}$$

The dissociative chemisorption of H_2 on the active sites was directly measured by Walker and co-workers [45] in their study of the strong retardation of the C-CO_2 reaction by small amounts of H_2. The isotherm of H_2 adsorption on two polycrystalline graphites (SP-1 and Graphon) in the temperature range investigated (950 to 1100°C) was indeed Langmuirian with dissociation of H_2 [45].

From the foregoing review on the C-H_2O reaction on multilayer edges at temperatures higher than above 800°C, it appears likely that mechanism C is operative and Eq. (25) is valid. However, more work on the multilayer edge rates is needed before a conclusion on the rate equation and mechanism can be reached.

B. Rate Equation and Mechanism on
    Single-Layer Edges

An etch-decoration TEM photograph for the C-H_2O reaction is shown in Fig. 14. All etch pits formed by H_2O in the temperature range studied (600 to 900°C) are hexagonal and all have the same orientation. The orientation was identified by matching the electron micrograph with its selected area electron diffraction pattern using TEM. The etch pits are composed of $\{10\bar{1}0\}$ faces, or the zigzag planes, on the six sides. The conformation and orientation of the pits have been interpreted in line with the foregoing discussion (discussion in Fig. 12). The detailed results of the reaction will be published elsewhere [46]. Portions of the kinetic results are summarized below.

The reaction has been studied in detail at temperatures of 700, 800, and 900°C. The partial pressures were 0 to 0.03 atm for H_2O and 0 to 0.06 for H_2. The turnover frequencies fit the following rate equation:

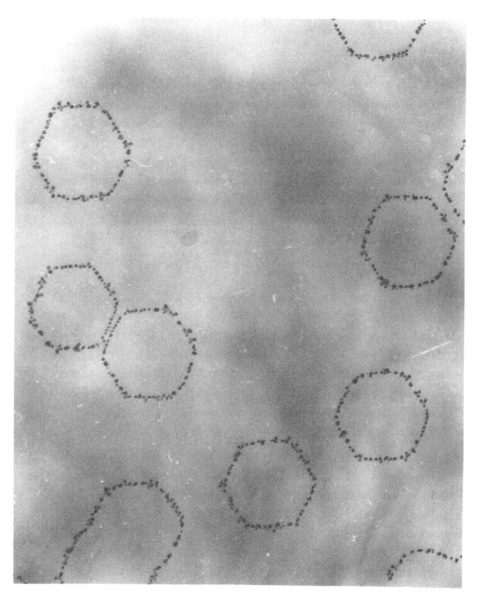

1 µm

FIG. 14 TEM micrograph of gold-decorated monolayer etch pits on the basal plane formed by 23 torr H_2O (in 1 atm N_2) at 600°C for 14 h. (From Ref. 24.)

$$\text{Rate} = \frac{k_1 P_{H_2O}}{1 + k_2 P_{H_2}^n + k_3 P_{H_2O}} \qquad (29)$$

where $n = 0.50$ at 700°C, $n = 1.0$ at 900°C. It is clear that the
reaction mechanism is a mixed one, with mechanism C becoming predominant at lower temperatures, and mechanism B predominant at
higher temperatures.

The rate constants at 800°C are $k_1 = 49.3$ atm^{-1} s^{-1}, $k_2 = 802$
atm^{-n}, and $k_3 = 6.17$ atm^{-1}. The temperature dependencies for k_1 and
k_3 are 75.6 and 35.2 kcal, respectively. Although it is not possible
to determine all elementary rate constants (i.e., the i and j values,
from the k values), the values of j_3 has been determined as k_1/k_3:

$$j_3 = 1.8 \times 10^9 e^{-40,400/RT} \qquad s^{-1} \qquad (30)$$

It is interesting to compare the rate constant j_3 for the elementary reaction $C(O) \rightarrow CO$ for the $C + H_2O$ reaction and the $C + CO_2$
reaction. The values of j_3 for the $C + CO_2$ reaction, given by Eq.
(4), are indeed surprisingly close to the values for the $C + H_2O$
reaction. Based on this comparison, the $C(O)$ complex is apparently
the same regardless of the oxygen precursor. The $C(O)$ complex on the
multilayer edges, as indicated by the values of j_{30} and E_3 in Table
1, is apparently different from that on the single-layer edges.

C. Kinetics of Vacancy Formation on the Basal Plane

The rate of abstraction of the basal-plane atoms (with three sp^2
bonds) has been measured in the same temperature and pressure ranges
as in the turnover frequency measurements [46]. The technique of
determining the rate of vacancy formation is by counting the smaller
pits, as detailed in the section on the $C-CO_2$ reaction.

The abstraction rates at a partial pressure of 0.03 atm of H_2O
are, in cm^{-2} s^{-1}: 2.5×10^2 at 700°C, 4.4×10^3 at 800°C, and $2.2 \times$
10^5 at 900°C. These rates were measured without the addition of
hydrogen. With the addition of H_2, the lowest concentration added
in our study being 0.02 atm, the rate diminished drastically by over
two orders of magnitude (i.e., to an almost unmeasurable level) [46].

The results above may be interpreted by the same mechanism as in the C-CO$_2$ reaction, via abstraction by oxygen atoms formed by thermal dissociation:

$$H_2O \rightleftarrows OH + H \rightleftarrows O + H_2 \tag{31}$$

The probability of removal of carbon upon each collision by atomic oxygen, which is assumed to be in equilibrium with H$_2$O via Eq. (31), is

$$\varepsilon = 5.7 \times 10^{-10} \ (900°C); \quad 1.2 \times 10^{-10} \ (800°C);$$
$$1.0 \times 10^{-10} \ (700°C) \tag{32}$$

The values of ε for the three gasification reactions, by O$_2$, CO$_2$, and H$_2$O, can now be compared. The values of ε for the C-CO$_2$ reaction are given by Eq. (10). The values of ε for the C-O$_2$ reaction can be calculated from data by Hennig [8] and Evans and Thomas [16]. At 840°C, the value of ε is 5×10^{-10} as calculated from their data. The values of ε for the three reactions are indeed very close, which indicates that atomic oxygen is probably the species responsible for creating vacancies on the surface, regardless of the precursor molecule for forming the oxygen atoms.

Based on the value of ε, it is also possible to calculate the energy of single vacancy formation on the basal plane [47]. The energies thus calculated are 4.18 eV (700°C) and 4.48 eV (800°C), which are in line with theoretical and experimental values (measured by other techniques) [48]. The fact that the added H$_2$ strongly inhibits the vacancy formation also supports the mechanism that the vacancies are formed by atomic oxygen.

It should be noted that a previous study was made by Montet and Myers on the C-H$_2$O reaction by the etch-decoration technique [49]. They reported rates for the reaction in the temperature range 850 to 1075°C at water vapor pressures below 0.022 atm. The water vapor was generated by saturating helium in a 30 volume percent sulfuric acid solution. Undoubtedly, the reactant gas also contained sulfur oxides, which are known to retard the C + O$_2$ reaction. It is not surprising that their rates are lower than those reported here, by approximately one-half at 600°C.

VI. RELATIVE RATES OF GASIFICATION ON SINGLE-LAYER EDGES

A. C-H$_2$ Reaction Rates

Another carbon gasification reaction of commercial interest is the C-H$_2$ reaction. The turnover frequencies of this reaction on single-layer edges have been measured in our laboratory using the etch-decoration technique [50]. The reactant gas, H$_2$ in He, both of high-purity grades, was further purified by passing through a bed of activated carbon at liquid nitrogen temperature. The graphite sample was chlorinated at 650°C to remove the impurities before purging with argon and performing the reaction, as described in the experimental section. The etch pits were circular, and the turnover frequencies for 1 atm hydrogen pressure were 0.091 s^{-1} at 850°C, 0.35 s^{-1} at 900°C, and 0.72 s^{-1} at 930°C.

B. Relative Rates of Gasification by Various Gases

The relative rates of carbon gasification, with no mass transfer limitation, by various gases were carefully estimated by Walker et al. [20]. This comparison was made based on literature data on various polycrystalline graphites (and a coal char for H$_2$/H$_2$O rates). Thus the relative rates were for gasification of multilayer edges. The ratio of rates, normalized to 800°C and 0.1 atm pressure, is:

$$\text{Rates with } O_2:H_2O:CO_2:H_2 = 10^5:3:1:3 \times 10^{-3} \tag{33}$$

The rates measured by the etch-decoration technique are for single-layer edges. A direct comparison of the rates with different gases on single-layer edges can be made for $O_2:H_2O:CO_2$ at 600°C and for $H_2O:CO_2:H_2$ at 900°C as follows. At 600°C and 0.03 atm:

$$\text{Rates with } O_2:H_2O:CO_2 = 22:3.4:1 \tag{34}$$

At 900°C and 1 atm:

Rates with $H_2O:CO_2:H_2 = 1.6:1:0.01$ (35)

Due to different apparent activation energies of the four reactions, Eqs. (34) and (35) should not be compared directly. Among the four reactions, the C-H_2 reaction has the highest apparent activation energy, which near 900°C is 71 kcal.

C. Relative Rates on Single-Layer
and Multilayer Edges

In the preceding sections, relative rates on single-layer and multilayer edges have been given for the C-O_2 and C-CO_2 reactions. Based on the results given in Eqs. (33), (34), and (35), the relative rates on the two types of edge atoms can be estimated for the C-H_2O and C-H_2 reactions. Thus an order-of-magnitude comparison is given below:

For the C-O_2 reaction:

$$\frac{\text{Rates on multilayer edges}}{\text{Rates on single-layer edges}} = 0(1)$$ (36)

For the C-CO_2, C-H_2O, and C-H_2 reactions:

$$\frac{\text{Rates on single-layer edges}}{\text{Rates on multilayer edges}} = 0(3-4)$$ (37)

It is interesting that the rates are substantially higher on the single-layer edges for the reactions with CO_2, H_2O, and H_2, whereas the opposite is seen for the reaction with O_2. The higher rates on single-layer edges for the CO_2 reaction have been attributed to the higher bond energy in C(O) on the single-layer edge.

If a stronger C(O) bond on the single-layer edge is also expected in the C-O_2 reaction, the lower rate on the single-layer edge for this reaction must be due to a much lower rate of formation of the surface oxide in the chemisorption step. In other words, the rate of formation of C(O) from O_2 on the multilayer edges is much higher than that on the single-layer edge. For the C-H_2O and C-H_2 reactions, a situation similar to the C-CO_2 reaction may exist.

VII. CONCLUSIONS

The etch-decoration electron microscopy technique serves as a unique tool for measuring the rates of gas-carbon reactions on well-defined active sites on graphite (i.e., single-layer edge atoms). The turnover frequency on a single-layer edge is very different from that on multilayer edges measured by other techniques. Hence the etch-decoration technique does not give overall rates for any carbon material.

The technique has been well developed. A step-by-step, easy-to-follow experimental procedure which has been used to obtain reproducible data is given in this chapter. The chapter summarizes the etch-decoration results on the reactions of carbon with O_2, H_2O, CO_2, and H_2 by earlier workers and more recent studies in the author's laboratory. These results include turnover frequencies and rate equations on the single-layer edge, elementary rate constants, elementary rate steps on the surface, and the kinetics and mechanism of vacancy formation on the basal plane. The technique has also been used to elucidate the mechanisms of the catalyzed reaction by non-channeling and nonpitting catalysts. The results on a single-layer edge have also been compared with results on multilayer edges, whenever possible, and it is shown that these comparisons can yield new insights into the reaction mechanism.

The etch-decoration technique has proven a useful tool for obtaining new insights and fundamental understanding of the surface processes for gasification of active sites on carbon. There is no doubt that the technique will offer more in the future. It is important, however, that the results from this technique are interpreted in conjunction with the multilayer-edge (or overall) rate results obtained by gravimetric or other means.

ACKNOWLEDGMENTS

The author is indebted to Dr. C. Wong (now at General Motors Research Laboratories) and to P. L. Cen and K. L. Yang for many of the results reported in this paper. Portions of the work were supported by the National Science Foundation under Grant CPE-8120569.

REFERENCES

1. J. M. Thomas, in *Chemistry and Physics of Carbon,* Vol. 1 (P. L. Walker, Jr., ed.), Marcel Dekker, New York, 1965, p. 121.

2. G. R. Hennig, in *Chemistry and Physics of Carbon,* Vol. 2 (P. L. Walker, Jr., ed.), Marcel Dekker, New York, 1966, p. 1.

3. E. L. Evans, R. J. M. Griffiths, and J. M. Thomas, *Science 171,* 174 (1971).

4. R. T. K. Baker, *Catal. Rev. Sci. Eng. 19*(2), 161 (1979).

5. D. W. McKee, in *Chemistry and Physics of Carbon,* Vol. 16 (P. L. Walker, Jr., and P. A. Thrower, eds.), Marcel Dekker, New York, 1981, p. 1.

6. P. A. Thrower, in *Chemistry and Physics of Carbon,* Vol. 5 (P. L. Walker, Jr., ed.), Marcel Dekker, New York, 1969, p. 217.

7. I. M. Dawson and E. A. C. Follett, *Proc. R. Soc. (Lond.) A274,* 386 (1963).

8. G. R. Hennig, *J. Chem. Phys. 40,* 2877 (1964).

9. C. Wong, Ph.D. thesis, Department of Chemical Engineering, State University of New York at Buffalo, Amherst, N.Y., 1983.

10. R. T. Yang and C. Wong, *J. Catal. 85,* 154 (1984).

11. E. E. G. Hughes, B. R. Williams, and J. M. Thomas, *Trans. Faraday Soc. 58,* 2011 (1962).

12. R. T. Yang and C. Wong, *Science 214,* 437 (1981).

13. R. T. Yang and C. Wong, *J. Chem. Phys. 75,* 4471 (1981).

14. F. S. Feates and P. S. Robinson, *Third Conference on Industrial Carbon and Graphite,* Society of Chemical Industry, London, 1971, p. 233.

15. F. S. Feates, *Trans. Faraday Soc. 64,* 3093 (1968).

16. E. L. Evans and J. M. Thomas, *Third Conference on Industrial Carbon and Graphite,* Society of Chemical Industry, London, 1971, p. 3.

17. P. L. Cen and R. T. Yang, unpublished results, State University of New York at Buffalo, Buffalo, N.Y., 1983.

18. R. T. Yang and C. Wong, *Rev. Sci. Instrum. 53,* 1488 (1982).

19. C. Wong and R. T. Yang, *Ind. Eng. Chem. Fundam. 22,* 380 (1983).

20. P. L. Walker, Jr., F. Rusinko, Jr., and L. G. Austin, in *Advances in Catalysis,* Vol. 11 (D. D. Eley, P. W. Selwood, and P. B. Weisz, eds.), Academic Press, New York, 1959, p. 133.

21. J. L. Johnson, in *Chemistry of Coal Utilization* (M. A. Elliott, ed.), Wiley-Interscience, New York, 1981, p. 1491.

22. J. F. Strange and P. L. Walker, Jr., *Carbon 14,* 345 (1976).

23. K. L. Yang and R. T. Yang, *AIChE J.,* in press.

24. R. T. Yang and C. Wong, *J. Catal. 82,* 245 (1983).

25. M. Mentser and S. Ergun, *U.S. Bur. Mines, Bull. 664,* 1973, p. 42.

26. S. Ergun, *J. Phys. Chem. 60,* 480 (1956).

27. K. W. Lewis, E. R. Gilliland, and G. T. McBride, *Ind. Eng. Chem. 41,* 1213 (1949).

28. P. L. Walker, Jr., L. G. Austin, and J. J. Tietjen, in *Chemistry and Physics of Carbon,* Vol. 1 (P. L. Walker, Jr., ed.), Marcel Dekker, New York, 1965, p. 327.

29. C. Wong and R. T. Yang, *Carbon 20,* 253 (1982).

30. H. M. C. Sosnovsky, *Phys. Chem. Solids 10,* 304 (1959).

31. R. Duan and R. T. Yang, *Chem. Eng. Sci.,* in press.

32. N. R. Laine, F. J. Vastola, and P. L. Walker, Jr., *J. Phys. Chem. 67,* 2030 (1963).

33. P. L. Cen and R. T. Yang, unpublished results, State University of New York at Buffalo, Buffalo, N.Y., 1983.

34. G. Hennig, *J. Inorg. Nucl. Chem. 24,* 1129 (1962).

35. D. W. McKee, *Carbon 8,* 623 (1970).

36. P. S. Harris, F. S. Feates, and B. G. Reuben, *Carbon 12,* 189 (1974).

37. R. T. K. Baker, *J. Catal. 78,* 473 (1982).

38. R. T. K. Baker and J. J. Chludzinski, *Carbon 19,* 75 (1981).

39. R. T. Yang and C. Wong, *AIChE J. 29,* 338 (1983).

40. R. T. Yang and C. Wong, *J. Catal. 85,* 154 (1984).

41. C. Wong, R. T. Yang, and B. L. Halpern, *J. Chem. Phys. 78,* 3325 (1983).

42. R. B. Levy and M. Boudart, *Science 181,* 547 (1973).

43. P. L. Cen and R. T. Yang, unpublished results, State University of New York at Buffalo, Buffalo, N.Y., 1983.

44. R. C. Giberson and J. P. Walker, *Carbon 3,* 521 (1966).

45. D. L. Biederman, A. J. Miles, F. J. Vastola, and P. L. Walker, Jr., *Carbon 14,* 351 (1976).

46. K. L. Yang and R. T. Yang, unpublished results, State University of New York at Buffalo, Buffalo, N.Y., 1983.

47. P. L. Cen and R. T. Yang, *Carbon,* in press.

48. P. A. Thrower and R. M. Mayer, *Phys. Status Solidi 47,* 11 (1978).

49. G. L. Montet and G. E. Myers, *Carbon 6,* 627 (1968).

50. P. L. Cen and R. T. Yang, unpublished results, State University of New York at Buffalo, Buffalo, N.Y., 1983.

4

Optical Properties of Anisotropic Carbon

R. A. FORREST,[*] *H. MARSH, and C. CORNFORD*[†]

Northern Carbon Research Laboratories
University of Newcastle upon Tyne
Newcastle upon Tyne, England

B. T. KELLY

Springfields Nuclear Power Development Laboratories
United Kingdom Atomic Energy Authority (Northern Division)
Springfields, Salwick, Preston, England

Also affiliated with: U.K.A.E.A., Harwell, Oxfordshire, England
Present affiliation: Integrated Geochemical Interpretation Ltd.,
Hallsannery, Bideford, Devon, England

I. INTRODUCTION

Optical microscopy is now used extensively to study the microtexture
of anisotropic carbons. This chapter presents a theory and reviews
experimental data describing the optical properties of the class of
materials known as anisotropic carbons. There are available a review
of the optical properties of graphite by Ergun [1] and a review of
diamond by Davies [2]. However, for anisotropic carbons of heat-
treatment temperature (HTT) below 2000 K, formed, for example, by
pyrolysis of coal tar or petroleum pitch or of compounds such as
polyvinyl chloride, neither of these treatments is applicable.
Structure in these anisotropic carbons is based on the graphite
lattice, in which the carbon atoms are predominantly hexagonally
linked by sp^2 trigonal bonds, but without three-dimensional ordering;
that is, the lamellae show parallel but disordered stacking. This
chapter describes the techniques used in reflection microscopy and
presents a phenomenological theory of optical reflectivity. Emphasis
is put on the orientation of the constituent lamellae of the anisotropic carbon as the lamellae emerge at a polished surface of carbon
and the direction of the vectors of the polarized light incident on
the polished surface.

The structure in anisotropic carbon is assumed to have the
same local symmetry as the graphite lattice, as discussed by Ergun
[1]. Reference should be made to this study because the nomenclature
of this chapter follows that of Ergun [1].

In Sec. II of this chapter the mechanism of formation of anisotropic carbon is discussed and reviewed, while Sec. III views the
anisotropic carbon as locally similar to the graphite lattice, but
with different optical parameters. In anisotropic carbon the orientations of local basal plane normals change both gradually and rapidly
throughout the carbon. Using this approach, theoretical equations for
the optical reflectivity of surfaces, mechanisms for generation of observed extinction contours, and reflection interference colors of
polished surfaces are developed. In Sec. IV anisotropic carbon is
considered as a collection of "adsorbing elements" and estimates of
the size and ordering of these elements are used to interpret experimental data for a wide range of such carbons. The theory allows the
determination of optical parameters for the absorbing elements them-
selves and their relative degree of order can be described by an
"anisotropy index."

   These concepts are developed for the first time and it is shown
that there is compatability between the two approaches of Secs. III
and IV. Sections V and VI review experimental results for some anisotropic carbons and discuss how the theories of Sec. III can be applied
to these data.

II. FORMATION OF ANISOTROPIC CARBON

A. Introduction

When an organic precursor is heated in an inert atmosphere (pyrolyzed
or carbonized) volatile matter is evolved. On cooling, there remains
a black or gray solid which is termed a carbon (a char or a coke).
If the carbon has been formed via a fluid phase during carbonization,
it is termed a graphitizable carbon because it can be graphitized on
heating to about 3000 K [3]. If no fluid phase is present during
carbonization, the carbon will be nongraphitizable and it is often
termed a char. The term "graphitic carbon" means that there is three-
dimensional order based on the graphite lattice present in the material after it has been heated to about 3000 K [4]. The graphitizable
carbons have a local anisotropic structure present even at the lowest
of heat-treatment temperatures (HTTs) of about 650 K and this aniso-

tropy can be shown by x-ray diffraction, electron diffraction, phase
contrast electron microscopy, and optical microscopy.

The formation of anisotropy has been attributed [4,5] to the
presence during carbonization of a liquid crystal phase. By a liquid
crystal is meant a material that has well-defined fluid properties,
that is, its viscosity is measurable $\sim 10^2$ to 10^{11} N s m^{-2} (10^3 to
10^{12} poise), but in which there is well-defined order on a length
scale of about 10^{-6} m (1 μm) as compared to the "statistical" order
that exists in liquids, which is on a length scale of about 10^{-9} m
(1 nm). In these liquid crystals, which are usually short-lived,
physical forces of attraction between lamellar molecules are sufficiently strong to impose long-range order on the molecules. However,
the pyrolysis chemistry continues and chemical cross-linking between
the molecules joins them together irreversibly and the material
becomes polymeric. This fluid polymeric system has been termed
"mesophase."

The detailed chemistry of these pyrolysis processes is extremely
complicated, as indicated by Fitzer et al. [6]. However, in general
terms the process of formation of anisotropic carbon can be summarized
as a progressive depolymerization (loss of functional groups and cleavage of carbon chains), synthesis (cyclization), and dehydrogenative
repolymerization (joining together of aromatic rings to produce large,
lamellar polycyclic aromatic structures).

The concept of formation of a liquid crystal phase as the means
of establishing structure or order between the isotropic fluid pitch
and the anisotropic mesophase is of paramount importance. This concept and that of the polymeric mesophase are extremely useful as pic-
torial models of structure and form the basis of the optical models
for anisotropic carbon.

B. Liquid Crystals [7,8]

The name "liquid crystal" may seem logically to be a misnomer, but
it is a useful expression to describe certain intermediate phases
formed by some organic molecular systems that do not pass directly
from a solid to a liquid. These liquid crystals have properties in
many ways intermediate between solid and liquid phases.

If the reversible phase change from an isotropic liquid to a
liquid crystal can be brought about by a lowering of temperature,
the system is termed "thermotropic." It is also possible to form
liquid crystals by dissolving molecules in a suitable solvent. In
this type of system the transition is usually brought about by changing the concentration of the solution; here the system is termed
"lyotropic."

During the transition from a crystal to a liquid, symmetry or
order is significantly reduced. Crystals possess translational symmetry. In liquids this translational symmetry is not present, although it is incorrect to describe liquids as totally disordered. A
hypothetical "snapshot" of a liquid on a molecular scale would show
local positional ordering similar to that in a crystal. This ordering
is transient and does not persist in time. Estimates suggest that
liquid molecules vibrate 10 to 100 times before the local structure
(order) changes appreciably. The order in a liquid is only short
range; it extends over a few molecular dimensions. This type of
ordering in a liquid is referred to as "statistical order."

Other types of molecular ordering, apart from positional ordering, are commonplace in physical systems. These types of molecular
ordering can be described as "orientational" ordering.

The literature describing liquid crystals is totally devoted to
systems composed of rodlike molecules. Platelike (lamellar) systems
are possible, but they had not been studied before their importance
in formation of anisotropic carbon was appreciated [4]. In fact,
the only mention in the liquid crystal literature that anisotropic
carbon is formed via liquid crystals is a footnote in a paper by
Saupe [7].

1. Classification and Terminology

a. Rodlike liquid crystals [8]. It is useful to describe initially
liquid crystals composed of rodlike molecules, since similar notations can be used for platelike (lamellar) liquid crystals. Such
organic molecules with the rodlike structure [e.g., PAA (p-azoxyani-
sole)] have two p-substituted aromatic rings rigidly linked by a
double bond.

b. *Nematic liquid crystals.* The nematic system can be represented
as a collection of rods of constant length in space. The direction
of the long axis is represented by a unit vector $\hat{\underline{a}}$. The important
properties of nematic liquid crystals are the following:

1. The positions of the centers of mass of the molecules are
    randomly distributed as in a liquid. There is no long-
    range order.

2. The long axes of the molecules (represented by $\hat{\underline{a}}$) are
    aligned on the average in a common direction in space which
    is termed the "director" $\hat{\underline{n}}$. The director $\hat{\underline{n}}$ indicates an
    axis of rotational symmetry for all macroscopic properties.

3. In an infinite nematic liquid crystal the orientation of
    the director $\hat{\underline{n}}$ is abitrary in space. This is not true for
    real systems because minor factors (e.g., the boundary of
    the system or external fields) will determine the orienta-
    tion of $\hat{\underline{n}}$.

4. There are an infinite number of twofold rotational axes
    perpendicular to $\hat{\underline{n}}$; therefore, it is impossible to distin-
    guish between states with director vector $\hat{\underline{n}}$ and $-\hat{\underline{n}}$.

5. Either the constituent molecule is "achiral" (identical
    with its mirror image) or the system is "racemic" (a 1:1
    mixture of the left and right forms).

c. *Cholesteric liquid crystals.* This type of system is similar to
the nematic one, but property 5 is not present. The molecules are
chiral and can exhibit a helical structure.

d. *Smectic liquid crystals.* Smectic liquid crystals have a layered
structure in which the spacing is well defined and can be measured
by x-ray diffraction. They contain more order than nematic liquid
crystals and if a compound exhibits both phases, the smectic phase
is always stable at a lower temperature than the nematic phase.
There are different types of smectic phases due to different forms
of order in the layers; the simplest smectic is smectic A. The important points about smectic A are:

(a) There is some translational symmetry. The layers repeat over a distance d which is approximately equal to the length of the molecule.

(b) There is no long-range order in any of the layers; therefore, the layers can be described as two-dimensional liquids.

(c) As with nematic liquid crystals, there is a preferential direction in space $\hat{\underset{\sim}{n}}$, which is also an axis of rotation.

(d) There is no distinction between systems with states $\hat{\underset{\sim}{n}}$ and $-\hat{\underset{\sim}{n}}$.

Nematic and smectic phases have well-defined symmetry properties which can be represented by the symbols $D_{\infty h}$ for nematics and D_{∞} for smectics in the Schoenflies notation.

e. Lamellar liquid crystals. Lamellar liquid crystals have a plate or disk-shaped unit as the basic building unit.

Studies of the molecular species present during the short fluid phase of pyrolysis show that molecular weights of the molecules vary with values between about 500 and 1000 amu [5,9,10]. Evidence from nuclear magnetic resonance (NMR) studies of pitches indicates that about 90 percent of the protons may be attached to aromatic rings [9]. This leads to a model of a "typical" molecule, shown in Fig. 1 as a disk about 2.5 nm in diameter. The C/H ratio is dependent on starting materials but is about 2.0 [11]. This indicates that several methyl groups may be present. Electron spin resonance (ESR) studies [12] indicate the presence of stable free radicals in the pyrolyzing system. This typical molecule is used in discussions of liquid crystals in the carbonaceous pyrolyzing system.

To distinguish between the rodlike and lamellar liquid crystal systems, a subscript is used on the conventional symbols. This is shown in Table 1. For rodlike liquid crystals (N_R and S_R) the molecules are represented by a rod labeled by the vector $\hat{\underset{\sim}{a}}$ along the long axis. For the lamellar liquid crystals (N_L and S_L) the molecules are represented by a disk-shaped lamellae which has a unit vector $\hat{\underset{\sim}{s}}$ situated at the center of mass with direction normal to the lamellar

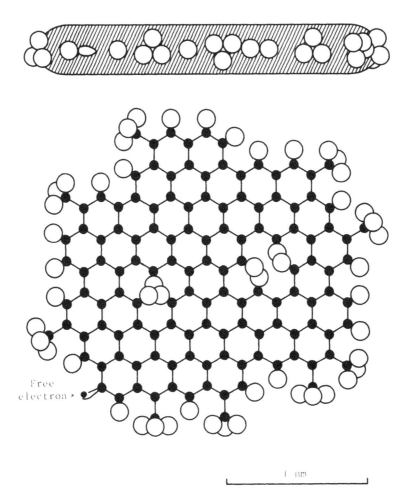

FIG. 1 Hypothetical molecule (amu ~1600) as may be found in mesophase. This model includes methyl groups, vacant sites in the aromatic lattice, and a broken bond (free electron). (From Ref. 31.)

surface. The N_L system has the same important properties (1 to 5
above) as listed for the nematic system with rodlike molecules (N_R).

It is important to note that in the N_L and S_L systems it is the
unit vector $\hat{s}$ (defining the normal to the molecule) which is aligned
in a direction n in space. The director vector $\hat{\underline{n}}$ is normal to the
molecular surface on which the molecules lie. Some early diagrams
of mesophase in systems give the impression that the projection of

TABLE 1   Conventional Symbols (N,S) for Liquid
Crystal Systems with Additional Suffices (R,L)
Indicating Shape of Constituent Molecule

Type of system	Constituent molecule	
	Rodlike	Lamellar
Nematic	N_R	N_L
Smectic	S_R	S_L

the lamellae, side on (giving a line), is analogous with the rod in
N_R and S_R systems. This is *not* correct because this does not have
the required rotational symmetry about the axis $\hat{n}$.

C.   The Fluid Phase of Pyrolysis

A pitch that is passing through the isotropic-anisotropic transition,
at a temperature of 630 to 700 K, is an extremely complex system. It
comprises molecules with a large spread of molecular weights inter-
acting with each other to polymerize via cross-linkages. Individual
molecules also change as volatile decomposition products are removed
in the gaseous phase, so creating free radicals. From this complex
fluid system, structural order arises, as detected in optical microscopy as a new optically active phase. It is believed that this
order is due to the formation of a transitory phase of lamellar
nematic liquid crystals [4,5]. Thus the anisotropic carbons owe
their origin to the facility of the pyrolyzing system to create
lamellar molecules (~1000 amu) which form initially a nematic lamellar liquid crystal which establishes the crystallographic order and
is carried through into the mesophase and carbon [13].

In the anisotropic graphitizing carbons the result of these
processes is to create planar macromolecules consisting of sheets of
hexagonally linked carbon atoms. These sheets are not perfect or
identical but contain holes, heteroatoms, and free electrons, as
shown in Fig. 2. In the fluid phase these planar molecules will
tend to be bound to each other surface to surface for relatively
long time periods by van der Waals dispersion forces, and the time
of bonding will increase (all other conditions, such as temperature,

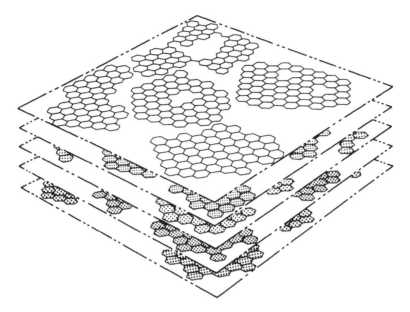

FIG. 2 Molecular structure of the carbonaceous mesophase (a N_L system) showing short-range stacking. (From Ref. 31.)

being constant) the larger these lamellar molecules become. Therefore, the liquid will appear locally to be fairly well ordered since
the normal to the molecules will be orientated preferentially in one
direction. However, this orientational ordering is effective only
over a short range, and if distances greater than about 5 to 10 nm
are considered, the correlation with orientation is lost. Thus,
even in the isotropic (as regards all optical parameters) fluid the
result of the van der Waals force is to line up the molecules surface
to surface. As the size of the molecules increases, a point is
reached when it is energetically favorable for the system to undergo
a phase transition and become a liquid crystal. Thus, from the
initial association of several molecules, there will grow the liquid
crystal and mesophase as seen by optical microscopy.

It has been shown that for an isotropic $\rightleftharpoons$ nematic transition
the phase transition must be first order [14]. This means that the
usual discontinuities in density and entropy and the presence of an

enthalpy of transition are to be expected. In conventional (N_R)
liquid crystals these have been observed but values are small and
the transition is described as weakly first order. Although it is
established [15] that the density of the mesophase is greater than
the density of the parent pitch, it is not known if there is a dis-
continuity at the phase transition itself.

 In pyrolyzing systems the onset of formation of liquid crystals
can be seen in optical microscopy by the formation of small (~1 μm),
optically anisotropic units. These anisotropic units can be studied
by cooling the pyrolyzing material to room temperature, embedding in
resin, and polishing, with examination of the polished surface by
optical microscopy. The use of hot-stage microscopy is also reported
[16,17].

 It was by the latter approach that Lewis showed that the formation of these anisotropic units is initially a physical rather than
a chemical process. Lewis [16] observed the anisotropic units as
spheres and once formed, the temperature was increased slightly,
causing the spheres to "dissolve" in the isotropic pitch. On lowering the temperature the spheres reappeared. This process could
be repeated for several cycles, but eventually the spheres could not
be dissolved in the heating cycle, indicating that chemical cross-
linking had progressed to establish a stable polymeric structure.

 The shape of anisotropic units is often spherical, but as shown
by Marsh et al. [18], there are many materials in which other shapes
are found. The spherical shape indicates a low viscosity of mesophase in pitch and has no importance other than that the shape of
the unit is determined by the minimum surface energy between unit
and fluid.

 During the early sixties Brooks and Taylor [9,19] showed both
by electron diffraction of thin sections and optical microscopy
(using reflected polarized light and crossed polars) on the cooled
material that the spheres had a well-defined morphology; they in-
voked the concept of liquid crystals. In the spheres the lamellar
molecules lie parallel to an equatorial section [4]. A review of

the structure in these spheres and of other structures found since
that date and a discussion of the evidence for the existence of the
nematic liquid crystal phase are available [20,21].

D. Formation of Bulk Mesophase

In this section the anisotropic units are assumed to be spheres,
because these usually are the shapes that coalesce easily. Once
spheres have been formed in the pitch, continued pyrolysis permits
their growth at the expense of the surrounding isotropic material.
When two spheres meet, coalescence is possible but does not always
occur, usually being dependent on the viscosity of the system. If
coalescence does occur, the resultant anisotropic unit may regain
its spherical appearance and may rearrange its internal structure
to eliminate many of the "defects" introduced by coalescence [17].
With continued coalescence the anisotropic phase of the system does
not retain its spherical shape but grows in extent until complete
conversion is attained. The resultant coalesced structure is referred to as "bulk mesophase" and no trace of original spheres can
usually be found. In the process of coalescence untransformed isotropic pitch or quinoline-insoluble particles such as carbon black
can be trapped in the anisotropic carbon. With increasing HTT the
viscosity of the mesophase increases until eventually it solidifies
to carbon. Once the system is solid, no further large-scale changes
in structure are possible [17].

The discussion above has established that:

1. The mechanism of establishment of crystallographic order in
    anisotropic carbon is via nematic lamellar liquid crystals.
2. The mesophase is essentially an irreversibly polymerized
    liquid crystal which exhibits significant fluidity.
3. During the growth and coalescence of the mesophase the
    lamellar stacking is disturbed such that a multiplicity of
    orientations result, manifesting itself as optical texture
    or microtexture as viewed by polarized light optical micro-
    scopy.

4. The lamella of the nematic liquid crystal system is essentially a polycyclic aromatic molecule, containing heteroatoms and joining defects (holes).
5. These aromatic lamellae are the molecular constituents
   stacked in anisotropic carbon.

The optical properties of the stacked lamellae presented at the
polished surface of carbon and the manner in which different HTTs
affect the perfection of stacking and of structure within individual
lamellae will be discussed in subsequent sections.

III. THE CONTINUUM THEORY OF OPTICAL PROPERTIES

A. Introduction

In Sec. II the concept of anisotropic carbon consisting of stacked
lamellar aromatic macromolecules, exhibiting significant variations
in stacking orientation, was attributed to the creation of a liquid
crystal phase followed by the solidification of the subsequent fluid
mesophase. The continuum of bent, twisted, or deformed stacking
arrangements based on the graphite lattice suggests that over small
volumes the local symmetry may be described in the same way as for
the graphite lattice (Figs. 3 and 4). This enables the theory for
crystallographic graphite to be applied to each of the volumes of
the paracrystalline anisotropic carbon.

It is important to note that graphite is a complicated material
on which to study optical properties. It is anisotropic (requires
the scalar optical parameters to be replaced by tensors) and is
strongly absorbing (optical parameters described by complex rather
than real numbers). In studies of reflectivity it is usually true
(but see Refs. 22 to 25) that light is reflected normally from a
polished surface. The situation of an absorbing uniaxial material
in which the optic axis is normal to the surface and light is reflected at an arbitrary angle is considered by Mosteller and Wooten
[26]; for anisotropic carbon it is normal incidence of light but
arbitrary angle of the optic axis to the surface which has to be
discussed.

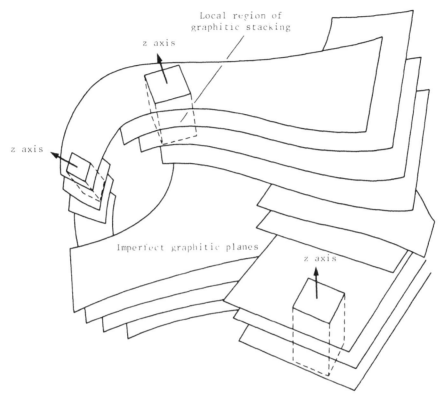

FIG. 3 Schematic diagram of anisotropic carbon showing twisted
imperfect graphitic planes with local regions of graphitic stacking.

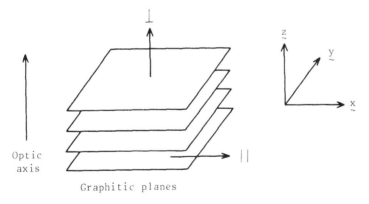

FIG. 4 Definition of the (x,y,z) coordinate system for a set of
graphitic planes. The directions parallel (∥) and perpendicular (⊥)
to the planes are shown.

B. Relevant Equations [27,28]

The equations in the chapter are written in gaussian units. This system of units allows the equations to be written in a simpler form than in MKSA units. In all discussions the materials are assumed to be nonmagnetic; thus the permeability $\mu = 1$.

The fundamental equations in all optical work are the Maxwell equations (see the list of symbols at the end of chapter).

$$\nabla \times H - \frac{1}{c} \frac{\partial D}{\partial t} = \frac{4\pi}{c} j \tag{1}$$

$$\nabla \times E + \frac{1}{c} \frac{\partial B}{\partial t} = 0 \tag{2}$$

$$\nabla \cdot D = 4\pi\rho \tag{3}$$

$$\nabla \cdot B = 0 \tag{4}$$

The properties of the material are specified by the constitutive equations. For isotropic materials these are

$$j = \sigma E \tag{5}$$

$$D = \varepsilon E \tag{6}$$

$$B = \mu H \tag{7}$$

Equation (5) is Ohm's law when E and j are time independent. When the field E is caused by an optical excitation of high frequency the conductivity σ is termed the "optical conductivity." For materials that are nonabsorbing the value of optical conductivity is vanishingly small. For anisotropic materials (only electrically anisotropic materials are considered in this chapter) the directions of D and E are not the same, as shown in the linear approximation by the tensor equation.

$$D = \varepsilon E \tag{8}$$

The dielectric tensor ε can be represented in a cartesian coordinate system by a matrix of nine numbers. These numbers are the components of the tensor and Eq. (8) can be rewritten in component form:

$$D_i = \varepsilon_{ij} E_j \tag{9}$$

It is possible to choose a particular coordinate system (the principal axes system) such that ε_{ij} is diagonalized. In this system there are only three independent components (instead of the six independent components in the general symmetric tensor). This principal axes system is extremely important; it is shown in Fig. 4 and is called the (x,y,z) system, where the z unit vector is in the direction of the optic axis. For a uniaxial material it can be shown that only two components are independent; these are perpendicular and parallel to the optic axis and are $\varepsilon_\perp$ and $\varepsilon_\parallel$, respectively.

When an absorbing material is considered, the optical conductivity is nonzero. Equation (5) can be rewritten for an anisotropic material:

$$\underset{\sim}{j} = \underset{=}{\sigma}\underset{\sim}{E} \tag{10}$$

$$j_i = \sigma_{ij}E_j \tag{11}$$

It is possible to diagonalize σ_{ij} by a suitable choice of coordinate system as was done with ξ_{ij}; however, it is *not* generally true that σ_{ij} and ξ_{ij} can be diagonalized simultaneously. This is possible for materials in which the symmetry is higher than or as high as orthorhombic. This simplification is possible in carbon since uniaxial symmetry is higher than orthorhombic.

In optical studies, wavelike solutions of the Maxwell equations are sought. The most important type of solution is a plane-wave solution where the field vectors $\underset{\sim}{E}$, $\underset{\sim}{B}$, $\underset{\sim}{D}$, and $\underset{\sim}{H}$ are proportional to a term $\exp[i(\underset{\sim}{k}\cdot\underset{\sim}{r} - \omega t)]$.

The wave vector $\underset{\sim}{k}$ has the same direction as the propagation of the wavefronts and its magnitude k is related to other quantities as follows:

$$k = \frac{2\pi}{\lambda} = \frac{n\omega}{c} = \frac{\omega}{v} \tag{12}$$

It is also convenient to define a vector $\underset{\sim}{n}$ such that

$$\underset{\sim}{k} = \frac{\omega}{c}\,\underset{\sim}{n} \tag{13}$$

$\underset{\sim}{n}$ is seen to be a generalization of refractive index; its direction is the direction of propagation of the wave and its magnitude is the value of refractive index in that particular direction.

This discussion of wave solutions is adequate for nonabsorbing materials, but for absorbing materials the notation must be extended. This can be done such that formally identical equations can be written for absorbing and nonabsorbing materials.

For a nonabsorbing material the amplitude of the wave remains constant. However, in an absorbing material the amplitude is damped since energy is lost from the electromagnetic field and is gained by the material generally in the form of heat. This fact can be expressed mathematically by making the wave vector complex, that is, by replacing $\underset{\sim}{k}$ by $\underset{\sim}{\hat{k}}$ in the exponential factor above:

$$\underset{\sim}{\hat{k}} = \underset{\sim}{k}_1 + i\underset{\sim}{k}_2 \tag{14}$$

$$\exp[i(\underset{\sim}{\hat{k}} \cdot \underset{\sim}{r} - \omega t)] = \exp[i(\underset{\sim}{k}_1 \cdot \underset{\sim}{r} - \omega t)]\exp(-\underset{\sim}{k}_2 \cdot \underset{\sim}{r}) \tag{15}$$

Equation (15) shows a wave which is decaying exponentially with distance. This type of wave is called an inhomogeneous wave and in general the directions of $\underset{\sim}{k}_1$ and $\underset{\sim}{k}_2$ are not the same. The quantities $\underset{\sim}{k}_1$ and $\underset{\sim}{k}_2$ are both physical; hence the vector $\underset{\sim}{\hat{k}}$ is a true complex quantity.

A corresponding equation to Eq. (13) can also be written for the absorbing case:

$$\underset{\sim}{\hat{k}} = \frac{\omega}{c} \underset{\sim}{\hat{n}} \tag{16}$$

All field vectors are proportional to a term $\exp\{i[(\omega/c)\underset{\sim}{\hat{n}} \cdot \underset{\sim}{r} - \omega t]\}$; hence the differential operators $\underset{\sim}{\nabla}$ and $\partial/\partial t$ can be replaced by simple multiplication.

$$\underset{\sim}{\nabla} \equiv \frac{i\omega}{c} \underset{\sim}{\hat{n}} \tag{17}$$

$$\frac{\partial}{\partial t} \equiv -i\omega \tag{18}$$

Replacing the differential operators in Eqs. (1) and (2), we have

$$\frac{i\omega}{c} \, \hat{\underset{\sim}{n}} \times \underset{\sim}{H} = \frac{4\pi}{c} \, \underset{\sim}{i} - \frac{i\omega}{c} \, \underset{\sim}{D} \tag{19}$$

$$\hat{\underset{\sim}{n}} \times \underset{\sim}{E} = \underset{\sim}{H} \tag{20}$$

Equation (19) can be simplified by introducing a complex vector $\hat{\underset{\sim}{D}}$ defined by Eq. (21):

$$\hat{\underset{\sim}{D}} = \underset{\sim}{D} + \frac{4\pi i}{\omega} \, \underset{\sim}{i} \tag{21}$$

giving

$$\hat{\underset{\sim}{n}} \times \underset{\sim}{H} = -\hat{\underset{\sim}{D}} \tag{22}$$

Another complex quantity, the complex dielectric tensor, can also be introduced as in Eq. (23), leading to a simple form of the constitutive Eq. (24):

$$\hat{\varepsilon}_{ij} = \varepsilon_{ij} + \frac{4\pi i}{\omega} \, \sigma_{ij} \tag{23}$$

$$\hat{D} = \hat{\varepsilon}_{ij} E_j \tag{24}$$

If Eqs. (23) and (24) are written in the principal axes system, both ε_{ij} and σ_{ij} are diagonal.

As emphasized by Wooten [28], $\hat{\underset{\sim}{D}}$ is a true complex quantity, that is,

$$\hat{\underset{\sim}{D}}(\text{physical}) = \text{Re}(\underset{\sim}{D}) + i \, \frac{4\pi}{\omega} \, \text{Re}(\underset{\sim}{i})$$

In the discussions of the various optical experiments in this chapter the light is always initially linearly polarized. Either of the vectors $\underset{\sim}{E}$ or $\underset{\sim}{H}$ can be used to define the state of polarization, but it is most convenient to use the vector $\underset{\sim}{E}$. A linearly polarized wave is one in which the direction of propagation $\underset{\sim}{k}$ and the electric vector $\underset{\sim}{E}$ both lie in a plane, the direction of which is fixed in space as the wave propagates.

For isotropic materials the amplitude of the reflected and transmitted (refracted) waves relative to the incident wave for any angle of incidence of light are described mathematically by the Fresnel reflection equations. For a nonabsorbing material the transmitted and reflected waves are found by multiplying the incident amplitude by a factor ρ. For the reflected wave, ρ is termed the reflection coefficient and it depends on the refractive index of the material n, the index of the surrounding medium n_0, and the angle of incidence. If E_i is the incident amplitude, the reflected amplitude E_r is given by

$$E_r = \rho E_i \tag{25}$$

All experimental observations are of intensities which are proportional to E^2; hence the quantity ρ^2 is actually measured. This is termed the reflectivity and has a value between zero and 1. It is also common to quote the percentage reflectivity R:

$$R = 100R = 100\rho^2 \tag{26}$$

For a material which is optically more dense than its surrounding medium, there is a phase change of π on reflection from this material. By convention the electric vector changes sign, while the magnetic vector has the same sign after reflection.

For an absorbing material the value of the reflection coefficient depends not only on the refractive indices n and n_0 but also on the absorptive properties of the material. The complex refractive index $\hat{n}$ can be written to show these absorptive properties:

$$\hat{n} = n(1 + i\kappa) = n + ik \tag{27}$$

When the equations for absorbing materials have n replaced by $\hat{n}$, the reflection coefficient ρ will also become complex. The physical meaning of a complex reflection coefficient is that there is a phase change on reflection in addition to the value of π mentioned above. This phase change is of great importance in discussions of reflection phenomena. The complex reflection coefficient $\hat{\rho}$ is defined by Eq. (28), where θ is the phase change on reflection:

$$\hat{\rho} = \rho e^{i\theta} \tag{28}$$

Most experiments are performed with air as the surrounding medium and consequently the value of n_0 is assumed to be 1.0. This simplifies many of the equations, but if a general medium is considered these equations must be extended by including n_0 explicitly.

The approach of Landau and Lifshitz [27] to the propagation of plane waves in an anisotropic material is extended in this chapter to include absorbing anisotropic materials. The basis of propagation of waves in any anisotropic material is the Fresnel equation. This is derived by eliminating $\underset{\sim}{H}$ from Eqs. (20) and (22):

$$\hat{D}_k = \hat{n}^2 E_k - \hat{n}_j E_j \hat{n}_k \tag{29}$$

Now $\hat{D}_k$ is eliminated by using $\hat{D}_k = \hat{\varepsilon}_{kj} E_j$, which yields

$$(\hat{n}^2 \delta_{kj} - \hat{n}_j \hat{n}_k - \hat{\varepsilon}_{kj}) E_j = 0$$

In order for this set of equations to be satisfied, it can be shown that the following is true:

$$\hat{n}^2 (\hat{\varepsilon}_x n_x^2 + \hat{\varepsilon}_y \hat{n}_y^2 + \hat{\varepsilon}_z \hat{n}_z^2) - [\hat{n}_x^2 \hat{\varepsilon}_x (\hat{\varepsilon}_y^2 + \hat{\varepsilon}_z) + \hat{n}_y^2 \hat{\varepsilon}_y (\hat{\varepsilon}_x + \hat{\varepsilon}_z) +$$
$$\hat{n}_z^2 \hat{\varepsilon}_z (\hat{\varepsilon}_x + \hat{\varepsilon}_y)] + \hat{\varepsilon}_x \hat{\varepsilon}_y \hat{\varepsilon}_z = 0$$

This can be simplified in uniaxial materials since $\hat{\varepsilon}_x = \hat{\varepsilon}_y \equiv \hat{\varepsilon}_\parallel$ and $\hat{\varepsilon}_z \equiv \hat{\varepsilon}_\perp$, giving

$$(\hat{n}^2 - \hat{\varepsilon}_\parallel)[\hat{\varepsilon}_\perp \hat{n}_z^2 + \hat{\varepsilon}_\parallel (\hat{n}_x^2 + \hat{n}_y^2) - \hat{\varepsilon}_\perp \hat{\varepsilon}_\parallel] = 0 \tag{30}$$

Equation (30) is the Fresnel equation for a uniaxial absorbing material, showing that there are two distinct values of $\hat{n}^2$:

$$\hat{n}^2 = \hat{\varepsilon}_\parallel \qquad \text{or} \qquad \frac{\hat{n}_z^2}{\hat{\varepsilon}_\parallel} + \frac{\hat{n}_x^2 + \hat{n}_y^2}{\hat{\varepsilon}_\perp} = 1 \tag{31}$$

It is necessary in crystal optics to define another vector $\hat{\underset{\sim}{s}}$, the ray vector, which is complex for absorbing materials. The ray vector has the same direction as the Poynting vector (proportional to $\underset{\sim}{E} \times \underset{\sim}{H}$) and indicates the direction of energy flow in the material. The magnitude of $\hat{\underset{\sim}{s}}$ is defined by

$$\hat{\underset{\sim}{n}} \cdot \hat{\underset{\sim}{s}} = 1 \tag{32}$$

Landau and Lifshitz [27] show that if equations are transformed according to the scheme shown in Eq. (33), the resulting equations are equally valid:

$$\left\{ \begin{array}{c} \underset{\sim}{E} \\ \underset{\sim}{\hat{n}} \\ \hat{\varepsilon}_k \end{array} \right\} \longleftrightarrow \left\{ \begin{array}{c} \underset{\sim}{\hat{D}} \\ \underset{\sim}{\hat{s}} \\ \hat{\varepsilon}_k^{-1} \end{array} \right\} \tag{33}$$

Applying Eq. (33) to Eq. (31), Eqs. (34) are derived:

$$\hat{s}^2 = \frac{1}{\hat{\varepsilon}_{\parallel}} \qquad \text{or} \qquad s_z^2 \hat{\varepsilon}_{\parallel} + (\hat{s}_x^2 + \hat{s}_y^2)\hat{\varepsilon}_{\perp} = 1 \tag{34}$$

Equations (31) and (34) indicate that for each value of $\hat{\underset{\sim}{n}}$ and $\hat{\underset{\sim}{s}}$ there are two values of $\hat{n}^2$ and $\hat{s}^2$. Physically, this means that two different waves propagate in each direction through the material.

C. Relationships Between Ordinary and
   Extraordinary Waves

The two waves referred to in the preceding section are called the ordinary and extraordinary waves. For the ordinary wave the solution $\hat{n}^2 = \hat{\varepsilon}_{\parallel}$ and $\hat{s}^2 = 1/\hat{\varepsilon}_{\parallel}$ is chosen and for the extraordinary wave the other solution in Eqs. (31) and (34) is used. For the ordinary wave it can be shown (Appendix A.1) that the vectors $\underset{\sim}{E}$ and $\underset{\sim}{D}$ are collinear and the propagation is exactly as if the material were absorbing and isotropic with a refractive index $\hat{n}_o = \hat{\varepsilon}_{\parallel}^{1/2}$.

The treatment of the extraordinary wave is more difficult, but simple forms of $\hat{n}_e^2$ and $\hat{s}_e^2$ can be found by defining $\hat{\theta}$ and $\hat{\theta}'$ as the angles (in general complex) between the vectors $\hat{\underset{\sim}{n}}$ and $\underset{\sim}{z}$ and $\hat{\underset{\sim}{s}}$ and $\underset{\sim}{z}$, respectively.

$$\frac{1}{\hat{n}_e^2} = \frac{\sin^2\hat{\theta}}{\hat{\varepsilon}_{\perp}} + \frac{\cos^2\hat{\theta}}{\hat{\varepsilon}_{\parallel}} \tag{35}$$

$$\frac{1}{\hat{s}_e^2} = \hat{\varepsilon}_{\perp}\sin^2\hat{\theta}' + \hat{\varepsilon}_{\parallel}\cos^2\hat{\theta}' \tag{36}$$

The vectors $\underset{\sim}{z}$ and $\hat{\underset{\sim}{n}}$ define a plane called the principal plane. It can be shown (Appendix A.2) that the set of vectors $\{\underset{\sim}{E}_e, \hat{\underset{\sim}{D}}_e, \underset{\sim}{H}_o, \hat{\underset{\sim}{n}}, \hat{\underset{\sim}{s}}\}$ lie in the principal plane and the set $\{\underset{\sim}{E}_o, \hat{\underset{\sim}{D}}_o, \underset{\sim}{H}_e\}$ are normal to the principal plane. It is important to be able to express the fields $\underset{\sim}{H}_o$ in terms of $\underset{\sim}{E}_o$ and $\underset{\sim}{H}_e$ in terms of $\underset{\sim}{E}_e$. Appendixes A.3 and A.4 show that

$$H_{\sim o} = \hat{\varepsilon}_{\parallel}^{1/2} \frac{\hat{k}_{\sim o} \times E_{\sim o}}{\hat{k}_o} \tag{37}$$

$$H_{\sim e} = \frac{1}{\hat{s}_e} \frac{\hat{s}_{\sim e} \times E_{\sim e}}{\hat{s}_e} \tag{38}$$

D.   Reflection at Normal Incidence

1.   *Derivation of Reflection Coefficients*

It is now possible to examine the problem of evaluating the reflection of monochromatic light at normal incidence from a polished surface of a uniaxial, homogeneous, absorbing, and nonmagnetic material where the orientation of the optic axis is at an arbitrary angle to the surface.

Several coordinate systems need to be defined. The principal axes system (x,y,z) is defined in Fig. 4. A new coordinate system is necessary to define the polished surface and the incoming light (the "microscope system"). The (η,ζ,ξ) system is a cartesian coordinate system such that η and ζ lie on the polished surface and the incoming light has its wave vector defined by $k_{\sim} = -k\xi_{\sim}$. The relative orientation of the two systems (x,y,z) and (η,ζ,ξ) is defined by the orientation angles Φ and Θ shown in Fig. 5. Only these two angles are necessary, due to the axial symmetry about $z_{\sim}$. The vectors $\xi_{\sim}$ and $z_{\sim}$ define the incident plane.

It is also necessary to define a two-dimensional coordinate system (u,v). This system lies in the η,ζ plane and is related to the (η,ζ) system by a rotation. The relationships between the (u,v) and the (η,ζ) systems are given by

$$\begin{pmatrix} u_{\sim} \\ v_{\sim} \end{pmatrix} = \begin{pmatrix} \cos\Theta & \sin\Theta \\ -\sin\Theta & \cos\Theta \end{pmatrix} \begin{pmatrix} \eta_{\sim} \\ \zeta_{\sim} \end{pmatrix} \tag{39}$$

For normal incidence of light on the surface the incident and principal planes are identical; therefore, the u direction specifies the direction of $E_{\sim e}$ and the v direction specifies the direction of $E_{\sim o}$.

Light incident normally on the surface with arbitrary linear polarization can be resolved into components with polarizations in the u and v directions. The response of the material to the ordinary

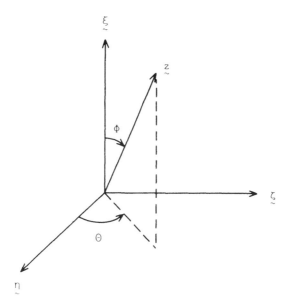

FIG. 5 The coordinate system (η, ζ, ξ) and the orientation angles Φ and Θ.

and extraordinary waves can therefore be found individually and these responses are then recombined to give the total response.

2. Reflection of Ordinary Wave

It is shown above that for the ordinary wave the material appears isotropic. Therefore, the standard derivation of reflection of electromagnetic waves from a surface is applicable. Appendix A.5 shows that the reflection coefficient $\hat{\rho}_o$ is given by $\rho_o e^{i\theta_o} = (\hat{n}_{\parallel} - 1)/(\hat{n}_{\parallel} + 1)$, where

$$\rho_o^2 = \frac{(n_{\parallel} - 1)^2 + k_{\parallel}^2}{(n_{\parallel} + 1)^2 + k_{\parallel}^2} \tag{40}$$

and

$$\tan \Theta_o = \frac{2k_{\parallel}}{n_{\parallel}^2 + k_{\parallel}^2 - 1} \tag{41}$$

3. Reflection of Extraordinary Wave

The derivation of the reflection coefficient for the extraordinary wave is more complicated and is given in Appendix A.6. $\hat{\rho}_e$ is of the same form as $\hat{\rho}_o$: that is,

$$\hat{\rho}_e = \frac{\hat{N} - 1}{\hat{N} + 1}$$

where

$$\hat{N} = \left(\frac{\cos^2\phi}{\hat{\varepsilon}_\parallel} + \frac{\sin^2\phi}{\hat{\varepsilon}_\perp} \right)^{-1/2} \tag{42}$$

It is thus possible to use the same expression for the reflection coefficient for both the ordinary and extraordinary waves:

$$\hat{\rho} = \frac{\hat{N} - 1}{\hat{N} + 1} \tag{43}$$

For the ordinary wave:

$$\hat{N} = \hat{n}_\parallel$$

For the extraordinary wave:

$$\frac{1}{\hat{N}^2} = \frac{\cos^2\phi}{\hat{n}_\parallel^2} + \frac{\sin^2\phi}{\hat{n}_\perp^2} \tag{44}$$

where

$$\hat{\varepsilon}_\parallel^{1/2} = \hat{n}_\parallel \qquad \text{and} \qquad \hat{\varepsilon}_\perp^{1/2} = \hat{n}_\perp \tag{45}$$

Equation (43) shows that the reflection coefficient for the extraordinary wave can be put into the standard form if the refractive index is given by Eq. (44). This is the same equation as that used for the refractive index when a wave is propagating through a uniaxial material. However, the angle ϕ has a different meaning for propagation and reflection. For propagation the angle is measured between the optic axis and the direction of propagation. For reflection the angle is measured between the optic axis and the normal to the surface.

4. Use of Reflection Coefficients to Calculate Reflectivity

The reflection coefficients derived above can now be used to calculate the intensity of light reflected from a polished surface. The initial state of the incoming light is linear polarization in the η direction. It can be shown (Appendixes A.7 and A.8) that the reflectivity is given by

$$R = \rho_\parallel^2 \ \sin^2\Theta + \rho^2(\Phi) \ \cos^2\Theta \tag{46}$$

where the notation $\hat{\rho}_o = \hat{\rho}_\parallel e^{i\Theta\parallel}$ is used for the ordinary wave reflection coefficient as it contains only "parallel" parameters $n_\parallel$ and $k_\parallel$, and $\hat{\rho}_e = \rho(\Phi)e^{i\Theta(\Phi)}$ for the extraordinary wave reflection coefficient since it depends on the orientation angle Φ.

Equation (46) is the final expression for reflectivity and shows that the phase changes on reflection are unimportant. The algebraic forms of $\rho^2(\Phi)$ and $\tan\Theta(\Phi)$ are extremely complicated and are not given explicitly in terms of Φ, $n_\parallel$, $k_\parallel$, $n_\perp$, and $k_\perp$.

5. Simplified Model to Calculate Reflectivity

The exact solution to the problem of reflectivity from a polished surface derived above is extremely complicated because of the form of $\rho(\Phi)$. For this reason a simplified model is derived which is algebraically more tractable than the exact solution. The validity and applicability of this model for carbons is then examined.

The simplifying assumptions of this model are:

1. The phase change $\theta(\Phi)$ has an identical value for all orientations. It is equated to $\theta_\perp$, which is analogous to $\theta_\parallel$, but $n_\perp$ and $k_\perp$ replace $n_\parallel$ and $k_\parallel$.

2. No analysis in terms of ordinary and extraordinary waves is undertaken but instead it is assumed that a "reflection response function" can be defined. In the principal axes system the response function is a diagonal matrix of the form

$$
\begin{pmatrix}
\hat{\rho}_\| & 0 & 0 \\
0 & \hat{\rho}_\| & 0 \\
0 & 0 & \hat{\rho}_\perp
\end{pmatrix}
$$

6. Geometry for the Model

The coordinate systems (x,y,z) and (η,ζ,ξ) are retained as before; the actual positions of the $\underset{\sim}{x}$ and $\underset{\sim}{y}$ axes are defined so that transformations between the two systems are possible. Figure 6 shows that the direction of $\underset{\sim}{y}$ is defined such that the projection of $\underset{\sim}{y}$ on the $\eta\zeta$ plane is in the same direction as the projection of $\underset{\sim}{z}$. Figure 6 shows how it is possible to express $\underset{\sim}{x}$, $\underset{\sim}{y}$, and $\underset{\sim}{z}$ in terms of $\underset{\sim}{\eta}$, $\underset{\sim}{\zeta}$, and $\underset{\sim}{\xi}$:

$$
\begin{pmatrix}
\underset{\sim}{x} \\
\underset{\sim}{y} \\
\underset{\sim}{z}
\end{pmatrix}
=
\begin{pmatrix}
\sin\Theta & -\cos\Theta & 0 \\
\cos\Phi\cos\Theta & \cos\Phi\sin\Theta & -\sin\Phi \\
\sin\Phi\cos\Theta & \sin\Phi\sin\Theta & \cos\Phi
\end{pmatrix}
\begin{pmatrix}
\underset{\sim}{\eta} \\
\underset{\sim}{\zeta} \\
\underset{\sim}{\xi}
\end{pmatrix}
\tag{47}
$$

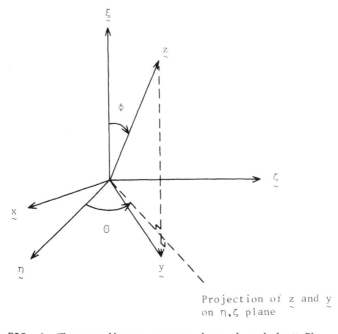

Projection of $\underset{\sim}{z}$ and $\underset{\sim}{y}$
on η,ζ plane

FIG. 6 The coordinate systems (x,y,z) and (η,ζ,ξ).

7. *Use of Response Function to*
    Obtain Reflectivity

In the principal axes system (x,y,z) the reflection response function
of the material is defined such that if $\Sigma_j e^{i\phi}\underset{\sim}{j}$ is the electric vector
of the incident wave, the electric vector of the reflected wave is
given by

$$\underset{\sim}{E}_r = E_j \rho_j e^{i(\phi+\theta j)}\underset{\sim}{j} \quad \text{where} \quad j \; \varepsilon \; \{x,y,z\} \tag{48}$$

Because of the uniaxial symmetry of the material (see Fig. 4),
the following identifications are possible:

$$\rho_x = \rho_y \equiv \rho_{\parallel} \qquad \text{and} \qquad \rho_z \equiv \rho_{\perp}$$

$$\theta_x \quad \theta_y \equiv \theta_{\parallel} \qquad \text{and} \qquad \theta_z \equiv \theta_{\perp}$$

Consider light linearly polarized in the η direction to be
incident normally on the polished surface. The electric vector of
the incident light is given by

$$\underset{\sim}{E}_i = Ee^{i\phi}\underset{\sim}{\eta} \tag{49}$$

Transforming to the principal axes system by the use of the inverse
of Eq. (47), we have

$$\underset{\sim}{E}_i = Ee^{i\phi}(\sin \Theta \underset{\sim}{x} + \cos \Phi \cos \Theta \underset{\sim}{y} + \sin \Phi \cos \Theta \underset{\sim}{z}) \tag{50}$$

Using Eqs. (48) and (50), the electric vector of the reflected wave
can be obtained:

$$\underset{\sim}{E}_r = Ee^{i\phi}(\sin \Theta \hat{\rho}_{\parallel}\underset{\sim}{x} + \cos \Phi \cos \Theta \hat{\rho}_{\parallel}\underset{\sim}{y} + \sin \Phi \cos \Theta \hat{\rho}_{\perp}\underset{\sim}{z}) \tag{51}$$

Transforming Eq. (51) back to the (η,ζ,ξ) system by use of Eq. (47)
yields

$$\begin{aligned}
\underset{\sim}{E}_r = Ee^{i\phi}[\underset{\sim}{\eta}((\sin^2 \Theta + \cos^2 \Theta \cos^2 \Phi)\hat{\rho}_{\parallel} + (\cos^2 \Theta \sin^2 \Phi)\hat{\rho}_{\perp}) \\
+ \underset{\sim}{\zeta}((-\cos \Theta \sin \Theta + \cos^2 \Phi \sin \Theta \cos \Theta)\hat{\rho}_{\parallel} \\
+ (\sin^2 \Phi \cos \Theta \sin \Theta)\hat{\rho}_{\perp} + \underset{\sim}{\xi}((-\sin \Phi \cos \Phi \cos \Theta)\hat{\rho}_{\parallel} \\
+ (\cos \Theta \sin \Phi \cos \Phi)\hat{\rho}_{\perp})]
\end{aligned} \tag{52}$$

The six geometric expressions in the brackets are termed g_1 to g_6. Equation (52) indicates that the reflected electric vector has a component of polarization in the ξ direction. This prediction is physically incorrect, as shown in the correct equation, Eq. (A23) of Appendix A.8. This component must be ignored when the intensity of reflected light is calculated because only light propagating normal to the surface will be detected at the eyepiece of the microscope. The electric vector at the detector is given by

$$\underset{\sim}{E}_r = Ee^{i\phi'}[\underset{\sim}{\eta}(g_1\rho_\|e^{i\theta_\|} + g_2\rho_\perp e^{i\theta_\perp}) + \underset{\sim}{\zeta}(g_3\rho_\|e^{i\theta_\|} + g_4\rho_\perp e^{i\theta_\perp})] \quad (53)$$

Using Eq. (A24) to calculate the reflectivity, we have

$$R = G_{12} + G_{34}$$

where

$$G_{ij} = g_i^2\rho_\|^2 + 2g_ig_j\rho_\|\rho_\perp \cos\psi + g_j^2\rho_\perp^2$$

and

$$\psi = \theta_\| - \theta_\perp$$

$$\Rightarrow R = \rho_\|^2 \sin^2\theta + (\rho_\perp^2 \sin^4\Phi + 2\rho_\|\rho_\perp \cos\psi \cos^2\Phi \sin^2\Phi + \rho_\|^2 \cos^4\Phi)\cos^2\theta \quad (54)$$

Equation (54) is of the same form as Eq. (46); however, the model solution has a simple algebraic form for $\rho^2(\Phi)$. Equation (54) also contains the phase difference on reflection ψ, which is not present explicitly in the exact equation (46).

In the derivation above it was noted that there is a component of light reflected parallel to the surface. This component is not used in deriving Eq. (54). However, if this light is included in the final reflected intensity (not actually possible physically), an additional term G_{56} must be included in the reflectivity. If this is done, the reflectivity is given by

$$R = G_{12} + G_{34} + G_{56}$$

$$\Rightarrow R = \rho_\|^2 \sin^2\theta + (\rho_\perp^2 \sin^2\Phi + \rho_\|^2 \cos^2\Phi) \cos^2\theta \quad (55)$$

Equation (55) has an even simpler algebraic form for $\rho^2(\Phi)$, having the advantage that the phase difference ψ vanishes. This equation, although it is not physically correct, is extremely useful becamse of its simplicity.

8. *Applicability of Model to Carbon*

The two model equations (54) and (55) for $\rho^2(\Phi)$ are good approximations to the exact form of $\rho^2(\Phi)$ calculated from Eqs. (43) and (44). Figure 7, a plot of reduced reflectivity and orientation angle, shows the results of calculations for a "standard" carbon: $n_{\parallel} = 2.0 + 1.663i$,

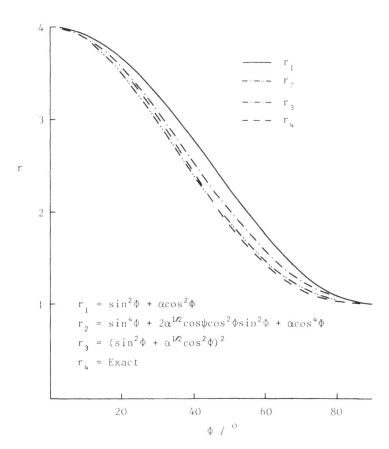

$$r_1 = \sin^2\Phi + \alpha\cos^2\Phi$$
$$r_2 = \sin^4\Phi + 2\alpha^{1/2}\cos\psi\cos^2\Phi\sin^2\Phi + \alpha\cos^4\Phi$$
$$r_3 = (\sin^2\Phi + \alpha^{1/2}\cos^2\Phi)^2$$
$$r_4 = \text{Exact}$$

FIG. 7 Values of reduced reflectivity (r) versus orientation angle (Φ) for the exact solution (r_4) and three model solutions (r_1 to r_3).

$n_{\perp} = 1.789 + 0.0i$, $\rho_{\parallel}^2 = 0.32$, $\rho_{\perp}^2 = 0.08$, and $\psi = 30°$. This carbon is used several times for calculations. The notations $\alpha = \rho_{\parallel}^2/\rho_{\perp}^2$ (anisotropy ratio) and $r = \rho^2(\Phi)/\rho_{\perp}^2$ (reduced reflectivity) are used.

Calculations are made using the incorrect model equation (55) (r_1), the correct model equation (54) (r_2), the correct model with $\psi = 0$ (r_3), and exact (r_4). The agreement between r_2 and r_4 is extremely good and even the simple model of Eq. (55) gives a reasonable approximation. A separate study for other values of optical parameters has shown this agreement to be consistently good for physical carbons. Equation (54) can thus be used with confidence for evaluating reflectivities.

9. *Effect of Rotation of Stage on Reflectivity*

It is found experimentally that if a particular location on the polished carbon surface is observed and the microscope stage is rotated relative to the direction of incident polarization, the measured reflectivity varies. This corresponds mathematically to the orientation angle Φ remaining constant while the rotation angle Θ is varied.

The anisotropy ratio α and the reduced reflectivity defined by Eq. (56) are used:

$$r(\Theta,\Phi) = \frac{F(\Theta,\Phi)}{\rho_{\perp}^2} \tag{56}$$

From Eq. (54) this gives

$$r = \alpha \sin^2\Theta + (\sin^4\Phi + 2\alpha^{1/2}\cos\psi\cos^2\Phi\sin^2\Phi + \alpha\cos^4\Phi)\cos^2\Theta \tag{57}$$

Plots of the reduced reflectivity are given in Fig. 8 for $\rho^2 = 0.08$ and $\alpha = 1, 2, 3$. For all three plots the value of ψ is taken as $30°$; this value is typical for actual experimental results [24].

For $\alpha = 1$ the material is not isotropic as might be expected, because the value of $k_{\parallel}$ is nonzero. This explains the deviation from an exact circle.

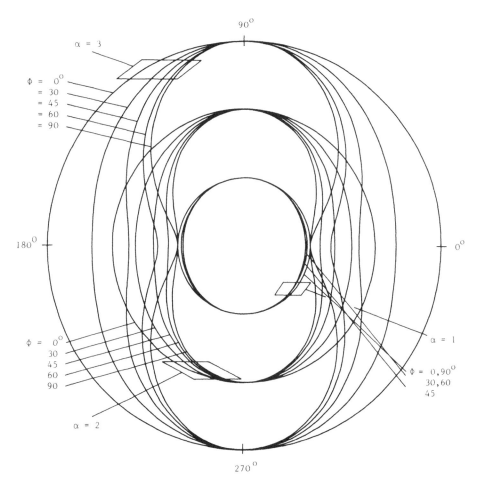

FIG. 8 Theoretical curves of reduced reflectivity (r) measured
radially as the sample is rotated (Θ varied), for different values
of orientation angle (Φ).

E. Generation of Extinction Contours

In studies of polished surfaces of anisotropic carbon, importance is
attached to the technique of using crossed polars in the reflectance
microscope. In this technique the light reflected from the surface
has to pass through the analyzer before reaching the microscope eyepiece. With the microscope set up as above, the surface of an anisotropic carbon shows black and white regions. It is also possible to

make quantitative reflectivity measurements with the analyzer present, as shown by Stevens [23]. The black regions are termed extinction contours, and examples are shown in Figs. 9 and 10. Extinction contours are extremely useful for structural analysis of carbons because, by rotation of the microscope stage, it is possible to map out the lamelliform structure of the carbon at the surface. This approach has been followed extensively by White [29-31].

The following discussion explains the generation of these extinction contours and shows the dependence of the extinction contours on the structure of the carbon. The approach of this derivation is similar to that of Woodrow et al. [32] but uses a different notation. The general case of the analyzer at an arbitrary angle (rather than 90°) is not discussed since it has not been required.

Equation (A23) gives the expression for the reflected electric vector. Because this has a component in the ζ direction as well as the original η direction, the light is now elliptically polarized.

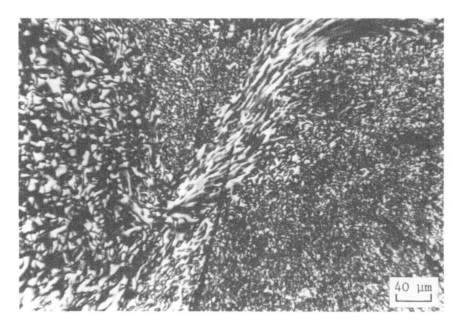

FIG. 9 Optical micrograph of polished surface of carbon prepared from Orgreave lean coal-tar pitch, HTT 800 K, 70 K h[-1].

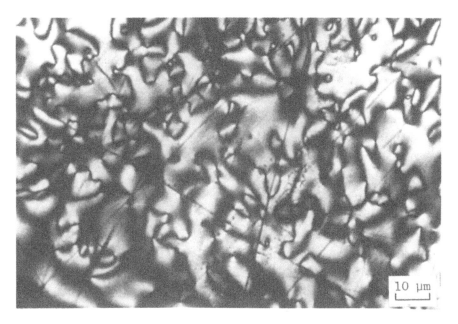

FIG. 10 Optical micrograph of polished surface of carbon prepared from D112 solvent-refined coal, HTT 800 K, 70 K h^{-1}.

If the analyzer is set to extinguish light polarized in the original η direction, only light polarized in the ζ direction will be detected. E_{ra}, the value of the electric vector after passing through the analyzer, is given by

$$\underset{\sim}{E}_{ra} = Ee^{i\phi'}(\cos \Theta \sin \Theta \rho(\Phi)e^{i\theta} - \cos \Theta \sin \Theta \rho_{\parallel}e^{i\theta_{\parallel}})\underset{\sim}{\zeta} \qquad (58)$$

From Eq. (A24) the intensity of the light reaching the detector can be given in terms of the intensity of the incident light I_0:

$$I = I_0 \cos^2 \Theta \sin^2 \Theta[\rho^2(\Phi) - 2\rho(\Phi)\rho_{\parallel} \cos \psi(\Phi) + \rho_{\parallel}^2] \qquad (59)$$

where

$$\psi(\Phi) = \Theta_{\parallel} - \Theta(\Phi) \qquad (60)$$

Equation (60) defines the phase change on reflection and this is seen to be a function of the orientation angle Φ. This is an

important difference from the simplified model, where the phase
change on reflection is assumed to have a constant value.

Equation (59) and the other derived reflectivity equations refer
to reflectivity of light from an extremely small region on the carbon
surface. If a region as large as that shown in Figs. 9 and 10 has to
be examined, Eq. (59) will be valid at each location of the carbon
surface. For most anisotropic carbons it is not unreasonable to
assume that the region being observed by the microscope is composed
of homogeneous material. If this is true, the values of $\hat{n}_\parallel$ and $\hat{n}_\perp$
are constant over the field of view. Therefore, the only parameters
in Eq. (59) to vary from point to point over the sample are the
angles Θ and Φ. Equation (59) can be used to calculate which values
of the orientation angles will give zero intensity. If $\Theta = m(\pi/2)$
(m is an integer), the intensity is zero. For a general value of Θ,
the intensity will also be zero if the factor in Eq. (59) that depends on Φ is zero. The exact solution shows, using Eq. (44), that
when $\Phi = 0$, $\hat{N} = \hat{n}_\parallel$ and consequently $\hat{\rho}(\Phi) = \rho_\parallel e^{i\theta_\parallel}$. This means that
the exact solution predicts a zero of intensity when $\Phi = 0$.

However, using the model solution, a zero of intensity will *not*
be found when $\Phi = 0$ because the phase difference ψ is independent
of orientation angle Φ, as mentioned above. The model solution is
therefore not useful in the case of crossed polars.

The meaning of these results is that no light will be detected
(and hence the photograph will be black) for any region that presents
a basal plane ($\Phi = 0$) to the surface, or has the projection of the
prismatic edges parallel or perpendicular to the incident polarized
light [$\Theta = m(\pi/2)$]. The distinction between the basal planes and
prismatic edges can be made by rotation of the carbon relative to
the incident polarized light. For basal planes the region will
remain dark on rotation, but for the prismatic edges the region
will become light and dark four times during a complete rotation.

The value of the intensity as Θ is varied will depend on the
orientation angle Φ. For our "standard carbon" the results of calculations of reflectivity with the crossed polars (R_+) are shown for
various orientation angles in Fig. 11.

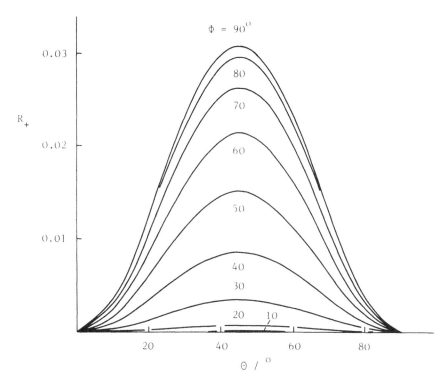

FIG. 11   Plots of reflectivity (R_+) versus rotation angle (Θ) for various orientation angles (Φ) with the polars crossed.

Figure 11 shows that for low values of Φ ("near basal plane") there is an extremely small maximum of intensity. Whereas the eye can easily detect the variation in intensity on rotation for high values of Φ ("near prismatic edge") this will not be true for the near-basal-plane case. It is probably true that for all values of $\Phi \lesssim 30°$ the eye will detect the region as isotropic on rotation. This could be very significant when regions are being picked out by eye for quantitative reflectivity measurements.

F.   Generation of Reflectance Interference Colors

1. *Introduction*

An experimental technique of considerable informative value in structural studies of carbon by optical microscopy is the generation of reflection interference colors. In this technique a phase retarder

plate is introduced into the optical system between the carbon specimen and the analyzer. With selected configurations of retarder plate
and polars it is possible to observe different regions of an anisotropic carbon surface in different colors.

Photographs reproducing these colors on carbon surfaces have
been published (e.g., Refs. 33 and 34). However, a full description
of the mechanism of generation of these colors and a standard terminology and notation for their description is needed [35].

2. Derivation of Intensity Equations

To produce these interference colors a reflection polarizing microscope is used, as in the generation of black-and-white extinction
contours. In addition to the analyzer a phase retarder plate, of
birefringent mica or quartz, is used. Either a half-wave plate or
a one-wave plate can be employed. Another coordinate system (the
"plate system") must be defined to describe the position of the
retarder plate relative to the microscope system (η, ζ, ξ) and the
principal axes system (x, y, z).

Because the plate is birefringent there are "fast" and "slow"
directions at right angles to each other and these are used to define
the plate system. Figure 12 defines the (f, s) system, which is derived from the (η, ζ) system by a rotation of $45°$. This angle is
always used experimentally and is therefore considered in this derivation rather than a general angle.

The relationship between the two systems is given by

$$\begin{pmatrix} \underset{\sim}{f} \\ \underset{\sim}{s} \end{pmatrix} = \frac{1}{\sqrt{2}} \begin{pmatrix} 1 & 1 \\ -1 & 1 \end{pmatrix} \begin{pmatrix} \underset{\sim}{\eta} \\ \underset{\sim}{\zeta} \end{pmatrix} \tag{61}$$

Unpolarized light from the source passes through the polarizer,
linearly polarized light strikes the surface, and elliptically
polarized light is reflected. This passes through the phase
retarder plate and the analyzer before detection at the eyepiece.

Equation (A23) gives the expression for the reflected electric
vector. This is rewritten in simplified form as

$$\underset{\sim}{E}_r = E e^{i\phi} (A \underset{\sim}{\eta} + B \underset{\sim}{\zeta}) \tag{62}$$

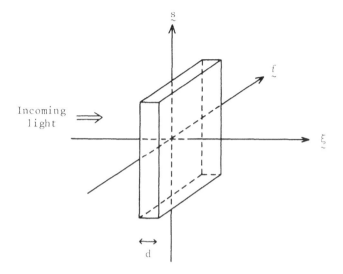

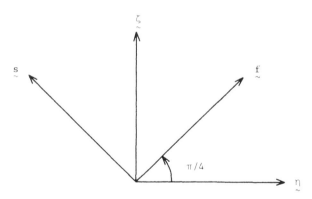

FIG. 12 Geometry of the phase retarder plate and the coordinate
system (f,s) relative to the (η,ζ) system.

The values of A and B can be seen in Eq. (A23). Using Eq. (61),
Eq. (62) is transformed into the (f,s) system:

$$\underset{\sim}{E}_r = \frac{Ee^{i\phi}}{\sqrt{2}} \, [(A + B)\underset{\sim}{f} - (A - B)\underset{\sim}{s}] \tag{63}$$

When light passes through any material there is a phase change which is dependent on thickness and refractive index and an amplitude change which is dependent on thickness and absorption coefficient. For the retarder plate the phase change on transmission depends on the polarization of the wave. The amplitude change on transmission can be assumed negligible because the plate is very transparent. These facts are shown by the transmission response function for the retarder plate. The incident light is given by Eq. (64) and the transmitted light by Eq. (65):

$$E_{\sim i} = Ee^{i\phi}\begin{pmatrix} a \\ b \end{pmatrix} \qquad \text{where} \qquad \begin{pmatrix} 1 \\ 0 \end{pmatrix} \equiv \underset{\sim}{f} \qquad \text{and} \qquad \begin{pmatrix} 0 \\ 1 \end{pmatrix} \equiv \underset{\sim}{s} \qquad (64)$$

$$E_{\sim t} = Ee^{i\phi}\begin{pmatrix} e^{i\theta_f} & 0 \\ 0 & e^{i\theta_s} \end{pmatrix}\begin{pmatrix} a \\ b \end{pmatrix} \qquad (65)$$

The 2×2 matrix in Eq. (65) is the transmission response function and if this is applied to Eq. (63), the electric vector is given by

$$E_{\sim rp} = \frac{Ee^{i\phi'}}{\sqrt{2}}\ [e^{i\theta_f}(A + B)\underset{\sim}{f} - e^{i\theta_s}(A - B)\underset{\sim}{s}] \qquad (66)$$

Transforming Eq. (66) by Eq. (61) into the (η, ζ) system gives us

$$E_{\sim rp} = \frac{Ee^{i\phi'}}{2}\ [\underset{\sim}{\eta}(Ce^{i\theta_f} + De^{i\theta_s}) + \underset{\sim}{\zeta}(Ce^{i\theta_f} - De^{i\theta_s})] \qquad (67)$$

where

$$C = A + B \qquad \text{and} \qquad D = A - B \qquad (68)$$

Consider now two positions of the analyzer in the optical system where:

1. The analyzer is set to extinguish light with incident polarization (crossed polars).
2. The analyzer is set to pass only light with the incident polarization (parallel polars).

After passing through the analyzer the electric vector is written as $E_{rpa\perp}$ in case 1 and as $E_{rpa\parallel}$ in case 2:

$$\underset{\sim}{E}_{rpa\perp} = \frac{Ee^{i\phi''}}{2}(Ce^{i\theta}f - De^{i\theta}s)\underset{\sim}{\zeta} \tag{69}$$

$$\underset{\sim}{E}_{rpa\parallel} = \frac{Ee^{i\phi''}}{2}(Ce^{i\theta}f + De^{i\theta}s)\underset{\sim}{\eta} \tag{70}$$

From Eqs. (A23), (62), and (68) the values of C and D are given explicitly in Eqs. (71) and (72):

$$C = \sin\Theta(\sin\Theta - \cos\Theta)\rho_{\parallel}e^{i\theta\parallel} + \cos\Theta(\sin\Theta + \cos\Theta)\rho(\Phi)e^{i\theta} \tag{71}$$

$$D = \sin\Theta(\sin\Theta + \cos\Theta)\rho_{\parallel}e^{i\theta\parallel} - \cos\Theta(\sin\Theta - \cos\Theta)\rho(\Phi)e^{i\theta} \tag{72}$$

Equation (A24) is used to evaluate the intensity for positions 1 and 2 from Eqs. (69) and (70):

$$I_{\perp} = \frac{I_0}{4}(CC^* + DD^* - CD^*e^{i\chi} - C^*De^{-i\chi}) \tag{73}$$

$$I_{\parallel} = \frac{I_0}{4}(CC^* + DD^* + CD^*e^{i\chi} + C^*De^{-i\chi}) \tag{74}$$

where

$$\chi = \theta_f - \theta_s \tag{75}$$

Here χ is the phase retardation of the plate. Equations (73) and (74) give the intensity for both settings of the analyzer and are the basis for understanding the generation of colors.

3. *Special Conditions*
Equations (73) and (74) can be simplified in certain special conditions as shown in Table 2.

G. Dispersive Properties
The dispersive properties of a material describe how the optical properties vary with wavelength. A simple treatment of the Lorentz model of absorption is given by Wooten [28], who shows that for an

TABLE 2 Optical Parameters and Intensities for Special Orientations of Anisotropic and Isotropic Materials

Type of material		Optical parameters and coefficients		Intensities
I	Isotropic	$\rho(\Phi) = \rho_\parallel \equiv \rho$ $\Theta(\Phi) = \Theta_\parallel \equiv \Theta$	$A = \rho e^{i\Theta},\ B = 0,$ $C = \mathcal{D} = A,\ AA^* = \rho^2$	$I_\perp = \dfrac{1}{2} I_0 AA^*(1 - \cos\chi)$ $I_\parallel = \dfrac{1}{2} I_0 AA^*(1 + \cos\chi)$
II(i)	Anisotropic material $\Theta = 0$	$A = \rho(\Phi)e^{i\Theta},\ B = 0,\ C = \mathcal{D} = A$		As above but $AA^* = \rho^2(\Phi)$ (depends on Φ)
II(ii)	$\Theta = 0, \dfrac{\pi}{2}$ $\Theta = \dfrac{\pi}{2}$	$A = \rho_\parallel e^{i\Theta_\parallel},\ B = 0,\ C = \mathcal{D} = A$		As above but $AA^* = \rho_\parallel^2$ (independent of Φ)
III(i)	Anisotropic material $\Theta = \pm\dfrac{\pi}{4}$ $\Theta = +\dfrac{\pi}{4}$ suffix$_+$	$A_+ = \dfrac{1}{2}(\rho_\parallel e^{i\Theta_\parallel} + \rho(\Phi)e^{i\Theta})$ $B_+ = \dfrac{1}{2}(-\rho_\parallel e^{i\Theta_\parallel} + \rho(\Phi)e^{i\Theta})$ $C_+ = \rho(\Phi)e^{i\Theta}$ $\mathcal{D}_+ = \rho_\parallel e^{i\Theta_\parallel}$	$C_+ C_+^* = \rho^2(\Phi)$ $\mathcal{D}_+ \mathcal{D}_+^* = \rho_\parallel^2$ $C_+ \mathcal{D}_+^* = \rho(\Phi)\rho_\parallel e^{-i\psi}$ $C_+^* \mathcal{D}_+ = \rho(\Phi)\rho_\parallel e^{i\psi}$	$I_{\perp+} = \dfrac{1}{4} I_0[\rho^2(\Phi) + \rho_\parallel^2 - 2\rho(\Phi)\rho_\parallel \cos(\chi - \psi)]$ $I_{\parallel+} = \dfrac{1}{4} I_0[\rho^2(\Phi) + \rho_\parallel^2 + 2\rho(\Phi)\rho_\parallel \cos(\chi - \psi)]$
III(ii)	$\Theta = -\dfrac{\pi}{4}$ suffix$_-$	$A_- = A_+$ $B_- = -B_+$ $C_- = \mathcal{D}_+$ $\mathcal{D}_- = C_+$	$C_- C_-^* = \mathcal{D}_+ \mathcal{D}_+^*$ $\mathcal{D}_- \mathcal{D}_-^* = C_+ C_+^*$ $C_- \mathcal{D}_-^* = C_+ \mathcal{D}_+$ $C_-^* \mathcal{D}_- = C_+ \mathcal{D}_+^*$	$I_{\perp-} = \dfrac{1}{4} I_0[\rho^2(\Phi) + \rho_\parallel^2 - 2\rho(\Phi)\rho_\parallel \cos(\chi + \psi)]$ $I_{\parallel-} = \dfrac{1}{4} I_0[\rho^2(\Phi) + \rho_\parallel^2 + 2\rho(\Phi)\rho_\parallel \cos(\chi + \psi)]$ or $I_{\perp-} = I_{\perp+}\{\chi \to -\chi\};\ I_{\parallel-} = I_{\parallel+}\{\chi \to -\chi\}$

isotropic material the complex dielectric function is given in terms of frequency by

$$\hat{\varepsilon} = 1 + \frac{4\pi Ne^2}{m} \frac{1}{(\omega_0^2 - \omega^2) - i\Gamma\omega} \tag{76}$$

Here N is the number of atoms per unit volume, ω_0 is the resonant frequency, and Γ (dimensions $[T]^{-1}$) represents the damping energy loss. Using Eq. (76), diagrams can be drawn for the variation of the real and imaginary parts of the dielectric function (Fig. 13), the real and imaginary parts of the refractive index (Fig. 14), and the reflectivity (Fig. 15) with frequency. These figures taken from Wooten are for $\hbar\omega_0 = 4$ eV, $\hbar\Gamma = 1$ eV, and $4\pi Ne^2/m = 60$ s^{-2}.

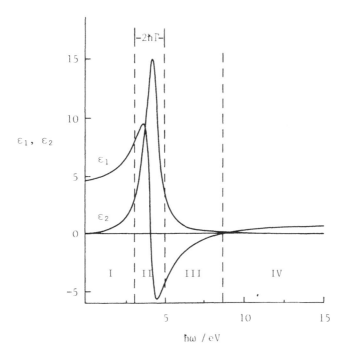

FIG. 13 Spectral dependence of ε_1 and ε_2. The curves are calculated for the case in which $\hbar\omega_0 = 4$ eV, $\hbar\Gamma = 1$ eV, and $4\pi Ne^2/m = 60$. The onset of region IV is defined by $\varepsilon_1 = 0$.

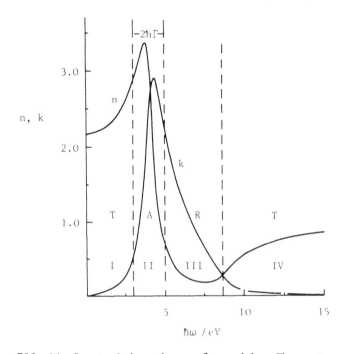

FIG. 14 Spectral dependence of n and k. The curves are calculated
from the values of ε_1 and ε_2 given in Fig. 13. The regions I, II,
III, and IV can be seen to be primarily transmitting (T), absorbing
(A), reflecting (R), and transmitting (T), respectively. These
results follow from consideration of Eq. (40) and the realization
that strong absorption takes place only in the neighborhood of a
transition frequency.

Four regions are labeled as primarily transparent (I and IV),
absorbing (II), and reflecting (III). To convert from the energy
scale of these figures to the more common wavelength scale $\lambda(nm)$ =
$1241/E(eV)$. For frequencies away from the absorption region $d\varepsilon_1/d\omega$
(and $dn/d\omega$) > 0, this is termed normal dispersion. In the region
of adsorption where ε_2 and ε_1 are appreciably larger than zero and
$d\varepsilon_1/d\omega$ (and $dn/d\omega$) < 0, this is termed anomalous dispersion. When
considered in terms of λ rather than ω, the signs in the inequalities
above are reversed. These trends are useful when studying variations
of optical parameters for carbons prepared under different experimental conditions, for example of heat-treatment temperature (HTT).

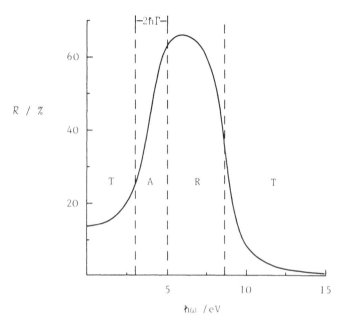

FIG. 15 Spectral dependence of reflectivity. The curve is calculated from the n and k values given in Fig. 14.

1. *Dispersive Properties of the Retarder Plate*

The retarder plate has been assumed to be completely transparent.
The entire visual spectrum is a region of normal dispersion and a
very simple algebraic expression can be given for $n(\lambda)$. This is
Cauchy's formula and is given in Eq. (77):

$$n(\lambda) = A + \frac{B}{\lambda^2} \tag{77}$$

Using Eq. (77), an expression for the dispersive properties of χ
can be found. Figure 12 shows the geometry of the retarder plate.
The phase retardation (χ) of the plate of thickness d is given by
Eq. (78), where the suffices f and s indicate the fast and slow
directions of the plate. Because of this definition, χ is always
negative.

$$\chi = \frac{2\pi d}{\lambda}[(A_f - A_s) + \frac{B_f - B_s}{\lambda^2}] \tag{78}$$

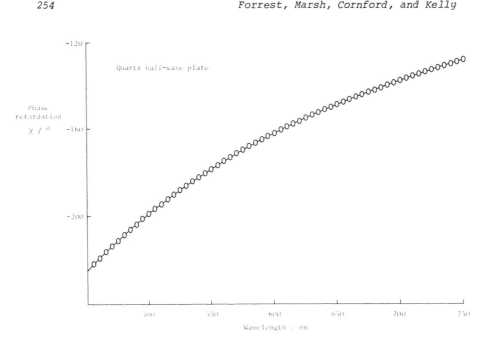

FIG. 16 Variation of phase retardation for a quartz half-wave plate
with wavelength. (From Ref. 36.)

Figure 16 shows the variation of χ with λ for a quartz half-wave
plate with λ_0 [$\chi(\lambda_0) = -\pi$] chosen to be 546 nm, calculated from
Eq. (78).

2. Dispersive Properties of Carbons

The general conclusion of analysis of these data is that anisotropic
carbons (cokes) with HTT $\gtrsim$ 870 K have a refractive index that increases with wavelength and an absorption coefficient that decreases
with wavelength over the visible spectrum. For cokes with HTT $\lesssim$ 870 K
the refractive index decreases with increasing wavelength and the absorption coefficient decreases with increasing wavelength. Typical
values for these variations are shown in Figs. 17 and 18.

For both of these sets of optical parameters the variation of
phase difference on reflection ($\psi = \theta_\parallel - \theta_\perp$) is calculated and results are shown in Fig. 19. For both normal and anomalous dispersion
$\psi > 0$ and $d\psi/d\lambda < 0$.

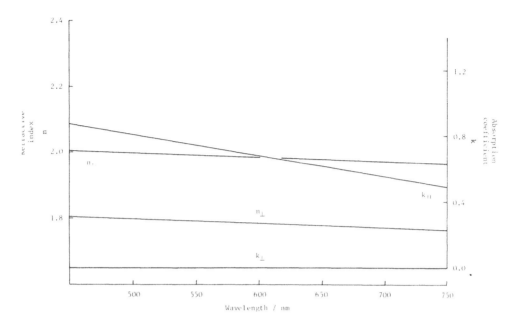

FIG. 17 Variation of optical parameters for carbons giving normal dispersion.

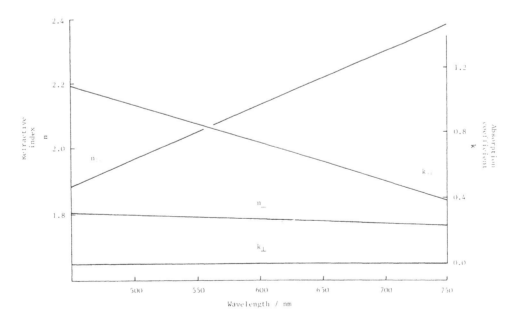

FIG. 18 Variation of optical parameters for carbons giving anomalous dispersion.

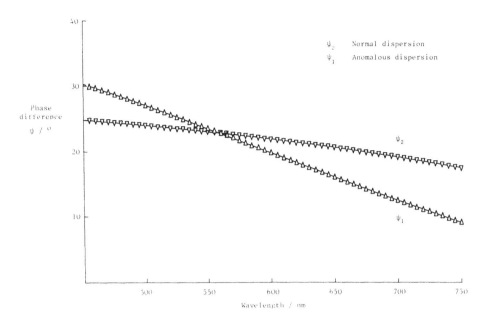

FIG. 19 Variation of phase difference (ψ) with wavelength for normal and anomalous dispersions.

3. Dispersive Properties of the Carbon/Retarder Plate System

For the half-wave plate λ_0 = 546 nm and $\chi(\lambda_0)$ = $-\pi$ and for the one-wave plate λ_0 = 546 nm and $\chi(\lambda_0)$ = -2π.

a. Isotropic material (I) or anisotropic material [II(i) and II(ii)] (Table 2).

$$\lambda/2 \text{ Plate:} \quad I_\perp(\lambda_0) = I_0 AA^* \qquad I_\parallel(\lambda_0) = 0$$

$$\lambda \text{ Plate:} \quad I_\perp(\lambda_0) = 0 \qquad I_\parallel(\lambda_0) = I_0 AA^*$$

This means that for a $\lambda/2$ plate and parallel polars and for a λ plate and perpendicular polars there is a zero of intensity at $\lambda = \lambda_0$. Therefore the middle (green) region of the spectrum is missing, giving a purple color.

b. Anisotropic material [III(i) and III(ii)] for $\lambda/2$ plate (Table 2). The parallel polars configuration is considered. Define a new intensity function by

$$\Delta I_{\|} = \frac{2}{I_0 \rho(\Phi)\rho_{\|}} (I_{\|+} - I_{\|-})$$ (79)

Appendix A.9 shows that $\Delta I_{\|}(\lambda_0 + \delta\lambda) < 0$ and $\Delta I_{\|}(\lambda_0 - \delta\lambda) > 0$. This enables the schematic diagram in Fig. 20 to be drawn. Figure 20 shows that for $\Theta = +\pi/4$ the minimum in the intensity spectrum is shifted from λ_0 to λ_1. This means that the high wavelength end of the spectrum (red) is lost and a resultant blue coloration will be seen. For $\Theta = -\pi/4$ the minimum in the intensity spectrum is shifted from λ_0 to λ_2. This means that the short wavelength end of the spectrum is lost and a coloration deficient in blue will be seen.

This analysis has demonstrated how the colors arise in the microscope, but it is not exact enough to predict actual colors. In fact, the coloration deficient in blue is experimentally found to be yellow.

c. Anisotropic material [III(i) and III(ii)] for λ plate (Table 2). The crossed polars configuration is considered. Defining a new intensity function by

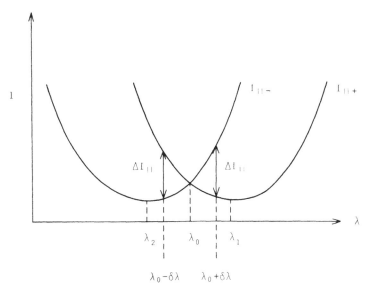

FIG. 20 Schematic diagram of the intensity spectra $I_{\|-}$ and $I_{\|+}$, showing the values of the intensity function $\Delta I_{\|}$ at $\lambda = \lambda_0 \pm \delta\lambda$.

$$\Delta I_\perp = \frac{2}{I_0 \rho(\Phi)\rho_\parallel} \, (I_{\perp+} - I_{\perp-}) \tag{80}$$

Table 2 shows that $\Delta I_\perp = -\Delta I_\parallel$. Using exactly the same argument as in Appendix A.9, we have

$$I_{\perp-} = I_{\parallel+} \qquad \text{and} \qquad I_{\perp+} = I_{\parallel-} \tag{81}$$

This means that for a λ plate the cases $\Theta = \pm\pi/4$ are reversed and the colors produced should be the reverse of the $\lambda/2$ plate.

H. Numerical Calculations of Intensity Spectra

In the preceding sections the intensity spectra have only been considered schematically, as in Fig. 20. In fact, only one point—the minimum intensity—is known with certainty. Therefore, numerical calculations have been made to give intensity spectra and related curves for different materials.

In the discussions above only the positions of the minimum of intensity are known and simplistic decisions are made on the actual color seen by the eye. In fact, the relationship between the intensity spectrum and the sensation of color seen by the eye is extremely complex, but it can be quantified in terms of chromaticity diagrams [37] not pursued here.

The human eye is not equally sensitive to light of all wavelengths, but is most sensitive to light in the green region of the visible spectrum. This differing sensitivity is expressed by the relative visibility curve $K(\lambda)$ [26]. This is found by plotting the reciprocal of the amount of energy at a wavelength λ needed to produce the same visible sensation as a reference wavelength λ_0. If λ_0 is taken at the eye's maximum sensitivity, then $0 < K(\lambda) \leq 1$. The shape of the $K(\lambda)$ curve is plotted in Fig. 21 for an average observer in bright light as in optical microscopy.

The product of intensity and relative visibility will be called the effective intensity, shown in Eq. (82):

$$J(\lambda) = I(\lambda)K(\lambda) \tag{82}$$

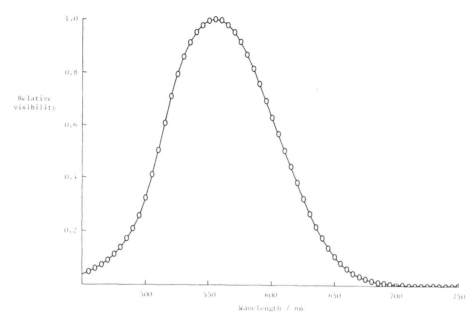

FIG. 21   Relative visibility factor $K(\lambda)$.

The effective intensity spectrum will give one or two peaks rather than the minimum found in the intensity spectrum, and these peaks give a reasonable indication of color.

1. Results of Numerical Calculations of Intensity Spectra

For initial calculations of intensity curves for anisotropic carbon the simplification of nondispersive carbon was made. By this it is meant that the optical parameters of the carbon do not vary with wavelength of light. The values of optical parameters used are for the standard carbon (i.e., $\rho_\perp^2 = 0.08$, $\rho_\parallel^2 = 0.32$). In order to write computer programs for carbon, Eq. (74) must be written explicitly. This is shown in Eq. (83):

$$R = \frac{1}{2}(s^2\rho_\parallel^2 + c^2\rho^2) + \frac{1}{2}\cos\chi[(s^2 - c^2)(s^2\rho_\parallel^2 - c^2\rho^2)]$$
$$+ sc\rho_\parallel\rho(\sin\psi\sin\chi + 2sc\cos\psi\cos\chi) \tag{83}$$

where the simplified notations $s \equiv \sin\Theta$, $c \equiv \cos\Theta$, and $\rho \equiv \rho(\Phi)$ have been used.

Using Eq. (83), the exact value of the reflectivity or the model value can be found depending on the choice for $\rho(\Phi)$. Many values of the reflectivity can be calculated for different values of the orientation angles Φ and Θ. Rather than reproduce many different plots it was decided that the point (λ_{min}, I_{min}) be plotted for each value of Φ, Θ. To do this the value of the minimum intensity was found for each Φ, Θ value and the corresponding λ value was then found. Figures 22 and 23 present this calculation for the nondispersive carbon.

Figure 20 shows that the curve of I against λ has a minimum value and for different orientations of the carbon relative to the polarized light (i.e., different Θ values) the minimum value (I_{min}) and corresponding wavelength (λ_{min}) vary. Figures 22 to 25 are plots of these minimum values for different values of Θ and Φ. Each curve is for one value of Φ and each point on that curve shows the point (I_{min}, λ_{min}) as the carbon is rotated (Θ varied) in 2° increments.

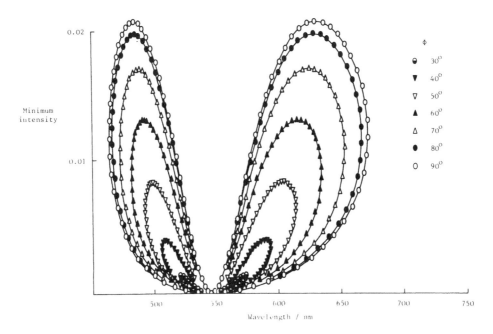

FIG. 22 Plots of (λ_{min}, I_{min}) for various values of orientation angles Θ and Φ for nondispersive carbon. Solution of Eq. (83) using the exact form of $\rho(\Phi)$.

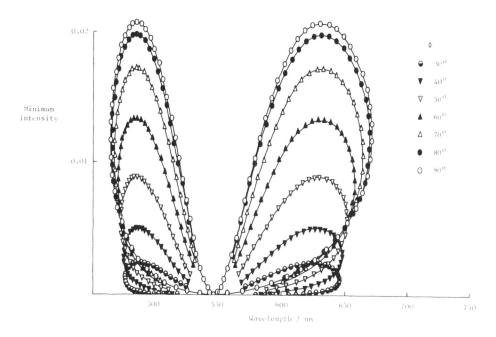

FIG. 23 Plots of (λ_{min}, I_{min}) for various values of orientation angles Θ and Φ for nondispersive carbon. Solution of Eq. (83) using the model form of $\rho(\Phi)$ given by Eq. (54).

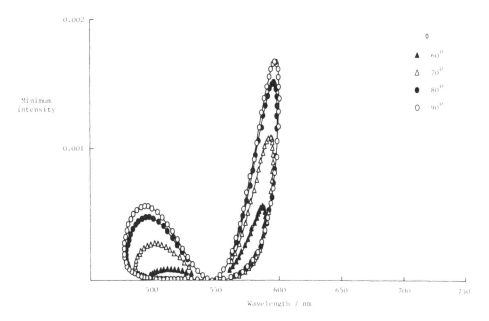

FIG. 24 Plots of (λ_{min}, I_{min}) for various values of orientation angles Θ and Φ for dispersive carbon. Solution of Eq. (83) using the exact form of $\rho(\Phi)$.

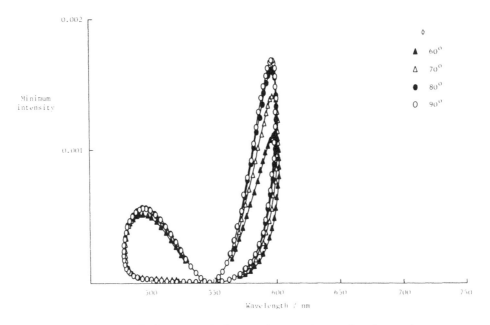

FIG. 25 Plots of (λ_{min}, I_{min}) for various values of orientation
angles Θ and Φ for dispersive carbon. Solution of Eq. (83) using
the model form of $\rho(\Phi)$ given by Eq. (54).

Considering when Φ = 90° in Fig. 22, the minimum intensity is at
546 nm when Θ = 0° but shifts towards longer wavelengths, along the
curve with smaller gradient as Θ increases. When Θ = 45° the minimum
is at 640 nm (corresponding to a blue color) and as Θ increases to
90° the minimum moves back to 546 nm (corresponding to a purple color).
As Θ increases to 135°, along the curve with larger gradient, the
minimum shifts towards shorter wavelengths and at Θ = 135° is at ~470
nm (corresponding to a yellow color). As Θ increases to 180° the
minimum moves back to 546 nm (corresponding to a purple color). The
same general trend is seen for the other values of Φ. But as the
configuration moves nearer and nearer to a description of basal plane
presentation the shift of the minimum from 546 nm as Θ varies becomes
less and less. The cases for Φ < 30° cannot be readily shown on the
same figure. This illustrates the important fact that the eye will
probably see all regions where $\Phi \lesssim$ 30° as purple and will therefore
interpret this region as isotropic. The eye is very insensitive to

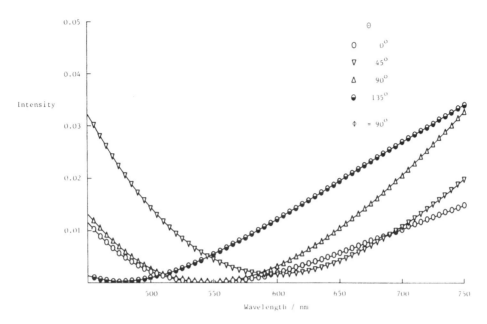

FIG. 26 Intensity spectra for the prismatic edge orientation for
various Θ values for dispersive carbon.

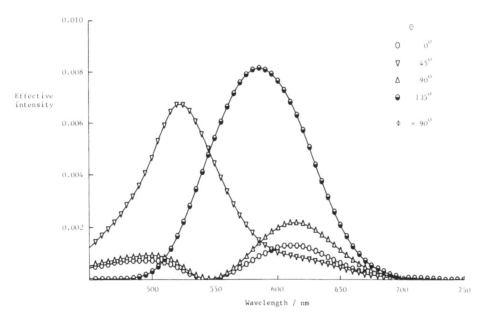

FIG. 27 Effective intensity spectra for the prismatic edge orientation for various Θ values for dispersive carbon.

small changes in orientation from a true basal plane and this must be remembered when regions are identified by eye in the microscope.

Figure 23 shows the same curves but for the model solution of Eq. (54). These show the same trends for large values of Φ but give incorrect predictions as $\Phi \rightarrow 0$ because of the inaccuracy of the model for low values of Φ. Preliminary studies using the incorrect model solution (Eq. (55)) appear to give better qualitative results for colors than the 'correct' model solution given by Eq. (54).

Similar calculations are made for a dispersive carbon. The values of the dispersive properties are the typical values shown in Fig. 18. Figure 24 gives the result for the exact solution and Fig. 25 gives the model solution using Eq. (54).

These figures show the same trends as in for nondispersive carbon. The shape of the curve is distorted compared with the non-dispersive situation. This is because of the variation of the optical properties with wavelength. For dispersive carbon four intensity spectra are calculated for a prismatic edge ($\Phi = 90°$), and the principal Θ values ($0°$, $45°$, $90°$, $135°$) are plotted in Fig. 26.

Corresponding to these intensity spectra the effective intensity spectra are calculated and given in Fig. 27. The effective intensity spectra show peaks at approximately the complementary color to the minima of the intensity spectra. The positions of the maxima indicate which color is seen.

Figure 27 shows that when $\Theta = 0°$ and $\Theta = 90°$, both curves have the same position of peaks but differ slightly in intensity. The two orientations have slightly different purple colors and experimentally it is possible to distinguish between these two purples with carbons. For $\Theta = 135°$ a peak is obtained in the yellow region of the spectrum (theoretical justification for the yellow color seen experimentally). For $\Theta = 45°$ a peak is obtained in the green region of the spectrum. In reality because of the sensitivity of the eye, this calculated method of judging colors by the effective spectrum is not absolutely correct. To agree with experiment, the eye in fact interprets this spectrum as blue. The results above can be summarized in the following rules for carbons.

2. *Summary of Bulk Geometry*

a. Parallel polars, λ/2 plate. The fast direction for the carbon is perpendicular to the layer planes, and the slow direction is parallel to the layer planes.

For the I_+ configuration:

Fast (sample) ‖ fast (plate) ⟶ blue   (84)

For the I_- configuration:

Fast (sample) ‖ slow (plate) ⟶ yellow   (85)

Relationships (84) and (85) are the rules for colors when using the λ/2 plate. When using these relationships with an optical microscope the direction of the incident light polarization (easily determined by the use of a separate marked polaroid sheet) must be found, as this is used as a reference (η axis) from which all angles and directions are measured.

b. Crossed polars, λ plate. Because of Eq. (81) it is immediately possible to write down the relationships between color and geometry.

Fast (sample) ‖ slow (plate) ⟶ blue   (86)

Fast (sample) ‖ fast (plate) ⟶ yellow   (87)

c. Effect of reversal of retarder plate. The retarder plate can be physically removed and reversed before returning it to the optical system. The effect of this is to interchange the directions of the fast and slow axes (i.e., $\underset{\sim}{f} \leftrightarrow \underset{\sim}{s}$). It can be shown that the effect of this is given by

$$I'_\perp = I_\perp \ \{C \leftrightarrow \mathcal{D}\}$$ (88)

$$I'_\parallel = I_\parallel \ \{C \leftrightarrow \mathcal{D}\}$$ (89)

Considering when $\Theta = \pm\pi/4$, Table 2 shows that

$$I'_{\parallel+} = I_{\parallel-}$$ (90)

$$I'_{\perp+} = I_{\perp-}$$ (91)

Equations (90) and (91) predict that reversing the retarder plate will change the colors seen such that blue ↔ yellow. This reversal of colors is found experimentally.

IV. THE DISCRETE THEORY OF OPTICAL PROPERTIES

A. Introduction

In the discussion above the carbon was viewed as small regions with
graphitic properties that were attached together in a composite form.
Rather than consider the graphitic planes as bending and twisting in
a continuous fashion through the material, an opposite viewpoint can
be taken to discuss these optical properties. Here the carbon may
be considered to be composed of discrete units—termed absorbing
elements—that are stacked with various degrees of order to give
the resultant carbon (in fact, they are not structually discrete).
This model uses data found from x-ray and electron microscope
studies. Electron microscope studies [38] show lattice images of
sizes as large as 100 nm, while x-ray diffraction data (e.g., Ref.
39) give values for L_a and L_c which are much smaller, of the order
of 5 to 10 nm. This is because x-ray diffraction demands better
order (i.e., planarity).

B. Absorbing Elements and Optical Properties

Using the data from these two types of study, an absorbing element
is defined as an imperfect graphitic sheet that contains within it
regions of higher order that would be detectable by x-ray diffraction. Figure 28 shows this schematically. The absorbing element

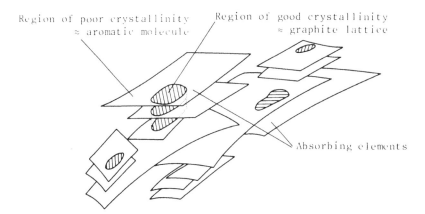

FIG. 28 Schematic diagram showing absorbing elements in anisotropic
carbons, containing regions with high crystallinity detectable by
x-ray diffraction.

has properties common to both the crystalline (x-ray diffraction) regions and aromatic (electron micrograph) region. In the former the electrons are free, as in graphite, whereas in the latter the electrons will be bound, as in aromatic molecules.

The optical properties of anisotropic carbons are dependent, basically, on the detail of structure within each absorbing element and also on the detail of modes of stacking of these units. To relate the observed value of reflectivity, R_{ob}, with the true value, R_τ, which refers to a perfectly stacked but not necessarily graphitic material, a function termed the anisotropy index (Ω) is introduced. Suppose that the anisotropic carbon is composed of two parts, a fraction Ω that is perfectly aligned and the remaining fraction $(1 - \Omega)$ that is not structured. The observed reflectivity can be represented by

$$R_{ob} = (1 - \Omega)\overline{R}_\tau + \Omega R_\tau \tag{92}$$

It is known that graphite has an immeasurably small absorption coefficient k perpendicular to the layer planes. Measurements with anisotropic carbons give a finite value for this quantity which could be due to the presence of less-ordered material. Therefore, measurements of the absorption coefficient can be used to give a definition of the anisotropy index Ω as in Eq. (93).

$$\Omega = \frac{k_{ob}^{max} - k_{ob}^{min}}{k_{ob}^{max}} \tag{93}$$

For perfectly stacked materials (e.g., graphite single crystals) Eq. (93) gives $\Omega = 1$, while isotropic materials (e.g., pitch) have $\Omega = 0$.

Equation (92) introduces the value of mean reflectivity ($\overline{R}$). Section IV.D gives expressions for various mean values and the result [Eq. (101)] is given here as Eq. (94):

$$\overline{R}_\tau = \frac{1}{3}(2\rho_\parallel^2 + \rho_\perp^2) = \frac{1}{3}(2R_\tau^{max} + R_\tau^{min}) \tag{94}$$

Using Eq. (92) for both observed maximum and minimum properties and Eq. (94), values of R_τ^{max} and R_τ^{min} are given by

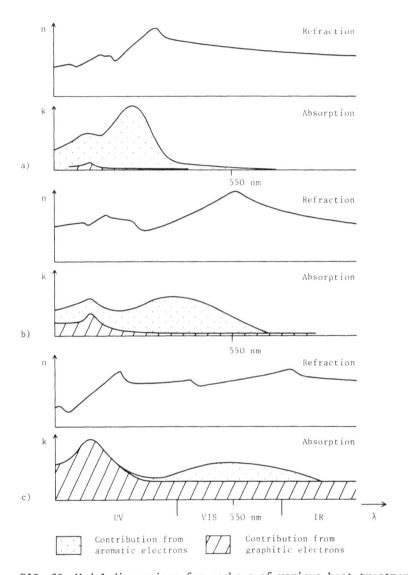

FIG. 29 Model dispersions for carbons of various heat-treatment
temperatures: (a) low temperature carbon (<700 K)—small-diameter
aromatic-type contribution predominates; (b) middle temperature
range (700 to 1100 K)—larger-diameter aromatic + some graphitic
contribution; (c) higher temperature carbon (>1100 K)—graphitic
contribution predominates.

$$R_\tau^{max} = \overline{R}_{ob} + \frac{1}{3}\frac{B_R}{\Omega} \tag{95}$$

$$R_\tau^{min} = \overline{R}_{ob} - \frac{2}{3}\frac{B_R}{\Omega} \tag{96}$$

where $\overline{R}_{ob} = (1/3)(2R_{ob}^{max} + R_{ob}^{min})$ is the observed mean value and B_R is the (observed) bireflection $R_{ob}^{max} - R_{ob}^{min}$.

It is suggested that the absorbing elements consist of lamellae in which the bonding is not totally graphitic because of heteroatoms and defects. There are, however, smaller regions where the bonding is more comparable to graphite.

Increasing heat-treatment temperature (HTT) will result in the bonding in the aromatic part of the sheet becoming more graphitic; that is, the graphitic contribution will increase at the expense of the aromatic contribution. This is shown in a very schematic and qualitative way in Fig. 29. Contributions to absorption as a function of wavelength are shown for three materials of increasing HTT. The absorption coefficient $k(\lambda)$ includes contributions from the graphitic and aromatic parts. As HTT increases the proportion of graphitic absorption increases; simultaneously the remaining aromatic absorption is shifted toward higher wavelengths. The latter effect is based on the observed shift in absorption properties of aromatic series (e.g., -acene and -phene series), where increasing molecular weight shifts the absorption maximum to higher wavelength. This has been termed the bathochromic shift [40]. The related dispersion curves for refractive index $n(\lambda)$ can be sketched into Fig. 29, the form following from Fig. 14, where an absorption maximum and a region of anomalous dispersion occur together.

C. Use of Bireflectance-Reflectivity
   Plots in Structure Determinations

Equations (95) and (96) refer to the parameter B_R, the bireflectance. Other B parameters can be defined by means of Eq. (97):

$$B_x = x^{max} - x^{min} \tag{97}$$

For $x = n$, k, the corresponding terms are bifringence and bisorbance.

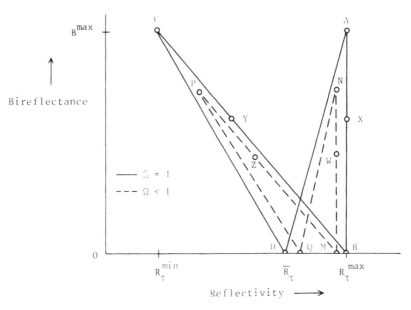

FIG. 30 Plot of bireflectance versus reflectivity.

It is also useful to define the normalized B parameter by division
by the maximum value (i.e., $b_x = B_x/x^{max}$). Corresponding to B_R is
b_R, the bireflectance ratio.

It is informative to present reflectivity data for carbons on
a B_R-R plot. Figure 30 shows diagramatically how the results can
be interpreted, and reference should be made to Table 3. The points

TABLE 3 Points Defined in Fig. 30 of B_R-R Plot
for Perfectly Aligned ($\Omega = 1$) Carbon

Point	Explanation	Θ	ϕ
A	Prismatic edge	$\frac{\pi}{2}$, $\frac{3\pi}{2}$	$\frac{\pi}{2}$
B	Basal plane	All	0
C	Prismatic edge	0, π	$\frac{\pi}{2}$
X	Intermediate orientation	$\frac{\pi}{2}$, $\frac{3\pi}{2}$	General
Y	Intermediate orientation	0, π	General
D	Mean reflectivity $\overline{R}$		

N, M, P, Q, W, Z correspond to the same geometrical orientation as A, B, C, D, X, Y, but for the former there is imperfect alignment (Ω < 1) compared to the perfect alignment for the latter points.

The horizontal distance between the point N and the line AB is a measure of Ω. The scatter of the various points N in a set of many reflectivity measurements is an indication of the variation of Ω throughout the material.

1. Derivation of Optical Parameters from Reflectivity Data

For completeness the established relationships for the optical parameters n and k of an isotropic material in terms of reflectivities are stated. Appendix A.10 also gives relationships between several optical parameters. The use of two media (usually air and oil) enables both optical parameters to be obtained. Equation (A31) of Appendix A.11 is rewritten here:

$$^{i}R = \frac{(n - n_i)^2 + k^2}{(n + n_i)^2 + k^2} \tag{98}$$

^{i}R is the reflectivity obtained in a medium of refractive index n_i. Following Ergun and McCartney [41], the parameter $T = (1 + R)/(1 - R)$ is calculated with

$$n = \frac{n^2_{oil} - 1}{2(^{oil}Tn_{oil} - {}^{air}T)} \qquad k^2 = 2n(^{air}T - 1) - (n - 1)^2 \tag{99}$$

D. Averages

Using the equations derived in Sec. III.D for reflectivity, it is possible to evaluate two important averages:

1. The Prismatic Average $<R>_p$

If a prismatic section is viewed and an average overall Θ value is taken, the prismatic average is obtained. This quantity is important since it is the value obtained for reflectivity if unpolarized light is reflected from a prismatic edge. For a prismatic edge $\Phi = \pi/2$ and $\rho(\pi/2)$ is known (= $\rho_\perp$); thus the exact form for reflectivity [Eq. (46)] can be used:

$$\langle R \rangle_p = \frac{1}{2\pi} \int_0^{2\pi} R(\Theta, \frac{\pi}{2}) \ d\Theta$$

$$= \frac{1}{2\pi} \int_0^{2\pi} (\rho_\parallel^2 \ \sin^2 \ \Theta + \rho_\perp^2 \ \cos^2 \ \Theta) \ d\Theta$$

$$= \frac{1}{2}(\rho_\parallel^2 + \rho_\perp^2) \tag{100}$$

If any other section ($\Phi \neq \pi/2$) is chosen, the average will depend on the orientation angle $[= (1/2)(\rho_\parallel^2 + \rho(\Phi)^2)]$.

2. The Three-Dimensional Average $\overline{R}$

The three-dimensional average $\overline{R}$ is obtained by averaging over all Θ and Φ values. It is the value for reflectivity from a random orientation of crystallites as discussed in Sec. III.B). The integral cannot be evaluated in closed form if the exact form for $\rho(\Phi)$ is used; thus the average is calculated for the two model forms [Eqs. (54) and (55)].

$$\overline{R} = \frac{1}{\text{area}} \int_{\text{sphere}} R(\Theta, \Phi) \ d\underset{\sim}{s}$$

$$= \frac{1}{4\pi} \int_0^{2\pi} \int_0^{\pi} (\rho_\parallel^2 \ \sin^2 \ \Theta + \rho^2(\Phi) \ \cos^2 \ \Theta) \ \sin \ \Phi \ d\Phi \ d\Theta$$

$$= \frac{1}{2}\rho_\parallel^2 + \frac{1}{4}\int_0^{\pi} \rho^2(\Phi) \ \sin \ \Phi \ d\Phi$$

1. Using Eq. (55) for $\rho^2(\Phi)$, the integration can easily be performed to give

$$\overline{R} = \frac{1}{3}(2\rho_\parallel^2 + \rho_\perp^2) \tag{101}$$

2. Using Eq. (54) for $\rho^2(\Phi)$, noting that the model assumes ψ independent of Φ, the integral gives

$$\overline{R} = \frac{1}{3}(1.8\rho_\parallel^2 + 0.8\rho_\perp^2 + 0.4 \ \cos \ \psi \rho_\parallel \rho_\perp) \tag{102}$$

Equation (101) is usually used in calculations since it has the advantages of simplicity and does not need an estimate of ψ.

E. Relationships Between Continuum and
    Discrete Theories

In the continuum theory there has been no provision made for imperfections in the material. Rather, these imperfections are implicitly
included because the values for the optical parameters are adjusted
to give the observed reflectivities. If the material were better
ordered, it would have different optical parameters. However, the
theory is capable of dealing with the fact that the local optic axis
of the material varies as a path of several-micrometer length is
followed through the material.

The information needed for a completely general treatment of
this curvature of space would be formidable; fortunately, there is
no need for such a detailed discussion. In the simplest example of
variation of optic axis direction, which (hopefully) is a slowly
varying function of distance (except at regions near a disclination
core), the variation can be described by two radii of curvature
shown in Fig. 31.

The simplest model assumes no variation of r_1 or r_2 with distance. Even this model is too complicated since what is required
is a way to reduce the curvature information to a single parameter
so that it can be compared with the anisotropy index Ω defined in
Sec. IV.B.

The details of the connection between the two theories are not
reported here, but the aim is to evaluate the average over a large
area of the reflectivity when the layer planes have a particular
radius of curvature. This average or observed value can then be
cast in the form of Eq. (92). In this way a relationship between
the anisotropy index (discrete theory) and the radius of curvature
(continuum theory) can be obtained.

V. REVIEW OF OPTICAL DATA

A. Cokes from Vitrinites

There are available comprehensive review articles giving optical
data for graphite [1], diamond [2], and coal [42-44]. However,
there is no comparable review of optical data for anisotropic

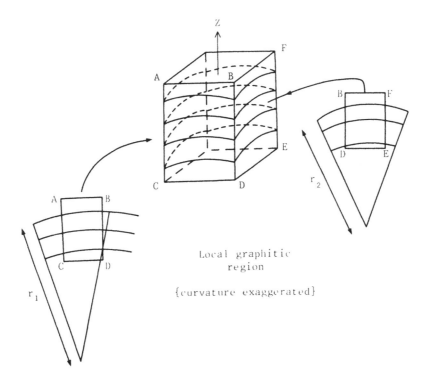

FIG. 31 Definitions of radii of curvature for graphitic sheets in local graphitic volume.

carbons. In this section results of reflectivity measurements on carbonized vitrinites are reviewed, and Sec. VI gives data for carbons derived from other precursors.

Only one paper, that of Marshall and Murchison [45], reports the dispersion data for vitrinites of different ranks and their corresponding semicokes. (Coals of 83.1, 88.6, and 93.1 percent C each heated to 450, 600, and 750°C.) Their results (Fig. 32) may be summarized as follows:

1. Absorption (Fig. 32b), initially showing an approximately flat dispersion appropriate to the coal rank, rises with increasing HTT initially at the blue end of the spectrum (450°C, 600°C) and then at the red end of the spectrum

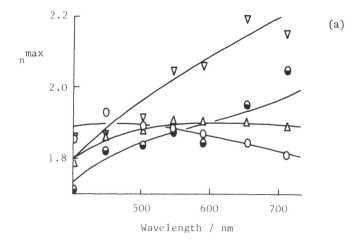

(a)

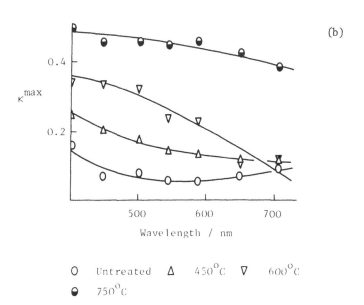

(b)

⭕ Untreated   △ 450°C   ▽ 600°C

◓ 750°C

FIG. 32 The dispersion at visible wavelengths of (a) refractive
index and (b) absorptive index of untreated vitrinite (88.6 percent
C) and its carbonization products. (Data from Ref. 45.)

(750°C), at which HTT it shows an approximately flat dispersion of a value between 0.4 and 0.5 for all three semicokes.

2. The refractive index dispersions (Fig. 32a), initially
gently falling (83.1 percent C), flat (88.6 percent C), or
rising (93.1 percent C) with wavelength, rise in value
especially at the red end of the spectrum, giving a markedly
positive trend with wavelength for cokes of HTT 600°C.
Values at this HTT were about 1.8 at the blue and 2.2 to
2.3 at the red end of the spectrum. At an HTT of 750°C the
values were lower again, in the region 1.8 to 2.0 and the
trend was much less positive.

Marshall and Murchison [45] interpreted these results in terms
of increased aromaticity giving rise to increased absorption and to
the negative slope in the refractive index curve. The fall in the
value of the refractive index, HTT 600 to 750°C, is thought to be
caused by structural reordering and is associated with the passing
of an absorption band, exactly as predicted by optical theory. Indeed, the fall in refractive index can be correlated with L_a, the
x-ray layer diameters, which indicate sizes of aromatic layers which
are large enough to give electronic absorption in the visible range
of the spectrum. The high slopes of both n and k dispersions indicate that extrapolation of data measured at an arbitrary wavelength
(e.g., 546 nm) to that of graphite is not valid. For example, semi-
coke from anthracite, HTT 600°C, shows a refractive index of 2.3,
higher than graphite at the red end of the spectrum, but shows values
of 1.85, about that of low volatile bituminous coal, in the blue.
The absolute values of the data reported by Marshall and Murchison
[45] must be heavily weighted toward the three-dimensional mean
(Sec. III.C), because average maximum measurements were used in the
calculation. For the fine-grained semicokes from the low rank
vitrains, measurements will have been made on a number of adjacent
mozaic units. The highest maximum will be the most significant
value, being a measurement made on the best aligned mozaics. No

data for minimum absorption were given, so no estimate of the anisotropy index can be made.

The trends of the n and k dispersions of all three vitrinite
semicokes can be viewed in terms of the dispersion relationships
depicted in Figs. 13, 14, 17, and 18. An interpretation of the
results could be that with increasing HTT the position of absorption
shifts to higher λ values, explaining the change in slope of both
the $n(\lambda)$ and $k(\lambda)$ results. The absolute values will depend on other
parameters, such as degree of anisotropy and physical density.

There are available further optical data all taken at single
wavelengths. The reflectivities of cokes, generally measured in oil,
are by Chandra and Bond [46,47], Chandra [48,49], Davis [50], Ghosh
[51], de Vries et al. [52], and Goodarzi and Murchison [53,54]. Only
Bond [47], de Vries et al. [52], and Goodarzi and Murchison [53,54]
determine refractive and absorptive indices. While various authors
have highlighted important carbonization factors such as particle
size and soak period (de Vries et al. [52]), pressure (Chandra [49]),
and heating rate (Ghosh [51]), the most detailed studies are those
of Goodarzi and Murchison [53,54]; see Fig. 33. Their results show
an initial rise in optical properties at 400 to 500°C from an initial
value which is dependent on the rank of the starting vitrinites (82.5,
88.0, 93.1 percent C). The absorption values rose for all three
vitrinites to values of 0.5 for cokes of HTT 950°C. In contrast,
the refractive index curves rose from values appropriate to the
starting vitrain rank, to reach maxima of about 2.0 (HTT 600°C).
These values then fall to about 1.9 (HTT 950°C) for all ranks of
starting materials.

Goodarzi and Murchison [53] interpreted the rise in adsorption
in terms of an increase in aromaticity and ordering with temperature,
not in terms of anomalous dispersion trends. They state that the
fall in the refractive index may correlate with the decrease in L_c
(crystallite height determined from x-ray line-broadening data),
which occurs over the same temperature range. They took the fall
in n and L_c for carbons of HTT about 600°C as indicative of a disordering or disruption effect.

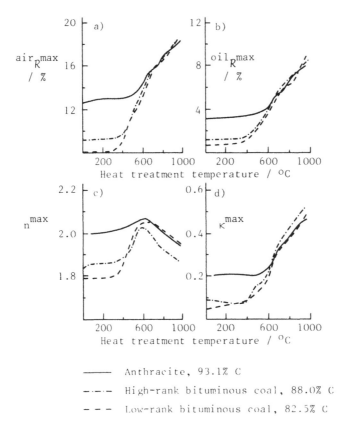

─────── Anthracite, 93.1% C

─··─··─ High-rank bituminous coal, 88.0% C

─ ─ ─ Low-rank bituminous coal, 82.5% C

FIG. 33 The variation of the maximum 246-nm values of (a) reflectivity in air, (b) reflectivity in oil, (c) refractive index, and (d) absorptive index of three vitrains with heat treatment. (From Refs. 53 and 54.)

It is probably preferable to account for the fall in the refractive index in terms of anomalous dispersion, because disordering would also affect the absorptive index and because it is in agreement with the slope of the dispersions discussed in Sec. III.G. It is thought that the crystallite diameter, L_a, rather than the crystallite height, L_c, has most effect on the optical properties, because L_a is related to the size of the planar aromatic units. Stacking, as such, would slightly affect the optical properties of individual layers. The absorptive and refractive indices are not independent variables, as the Kramers-Kronig dispersion relationship shows, and hence should not be interpreted independently. It is clear from the opposite trends of n and k (HTT > 600°C) that detailed interpretations of reflectivity, a complicated function of both, must be approached with caution.

In conclusion, data are in agreement with the model of carbonization (Sec. II) and the expected optical effects thereof (Sec. IV). Both the absorption and refraction agree with the migration of an absorption edge, from short to long wavelengths in carbons of increasing HTT treatment, this indicating an increase in size of conjugated aromatic units with temperature.

VI. EXPERIMENTAL STUDIES

A. Materials Used (Part 1)

In this section data are given for reflectivity studies of a wide range of anisotropic carbons. From these data it is possible to extract optical parameters and to apply the ideas of Sec. IV to obtain the true values of optical parameters and the anisotropy index. The variation of reflectivity with microscope stage rotation is investigated for comparison with Sec. III.

1. Orgreave Lean Coal-Tar Pitch

Carbons were prepared by carbonizing under nitrogen in open boats at atmospheric pressure, HTT 650 to 1200 K at 5 K min^{-1} with zero or 16 min of soak time.

2. Polyvinyl Chloride

Carbons were prepared by carbonizing under nitrogen in open boats at atmospheric pressure, HTT 650 to 1200 K at 5 K min^{-1} with 30 min of soak time. A carbon of HTT 1573 K was prepared under argon in a graphite resistance furnace.

3. Acenaphthylene

Carbons were prepared by carbonizing acanaphthylene in gold tubes at a pressure of 67 MPa, HTT 850 to 950 K at 5 K min^{-1} with zero soak time.

B. Sample Preparation and Microscopy

Carbons were mounted in a thermosetting resin. After curing and cutting with a diamond wheel to expose the carbon, the surface was polished with carborundum paper. The final polish was given using three grades of alumina powder on a Selvyt cloth lap.

Studies were made using the Vickers M41/M74 microscope and a Leitz (Orthoplan) microscope with photometer. The detail of optical measurement for variation of reflectivity with Θ is as follows. An area of specimen is located larger in area than that of the measuring aperture and which showed a bireflectance approaching that of the maximum. The polish has to be excellent and the surface precisely perpendicular to the optic axis of the microscope. The coordinates of these areas are noted and the areas are sketched. Reflectivity measurements are made in air and oil from identical areas, for every 10° or 20° intervals of stage rotation. A diamond prism, primary standard (^{air}R = 17.3 percent; ^{oil}R = 5.3 percent), is used both before and after the rotating procedures.

To assess the variation of reflectivity with HTT, reflectivities are taken from about 20 areas of a carbon sample. Areas are selected, using parallel polarizers and a half-wave plate, to exhibit isochromatic regions of good polish. Both basal and prismatic areas are chosen. Unless otherwise stated, all reflectivities are made at λ = 546 nm.

C. Results (Part 1)

Figures 34 and 35 show data for the anisotropic carbons as a function of HTT in air and oil. Figures 36 and 37 show reflectivity data for Orgreave lean coal-tar pitch HTT 730 K measured on the mesophase spheres as a function of sphere diameter. There is considerable scatter among the points (to be expected since the size of the spheres is uncertain because they will not all be equatorial sections) and the best straight line is drawn through the points. For perfect correlation between two variables the product moment correlation coefficient r = 1; for no correlation r = 0.

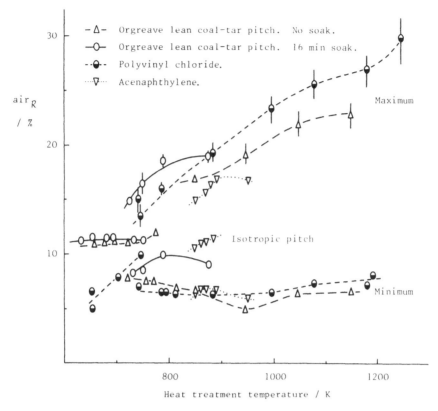

FIG. 34 Reflectivity data in air for anisotropic carbons of various
heat treatment temperature values.

Figures 38 and 39 present reflectivity data for Orgreave lean
coal-tar pitch carbon HTT 730 K as bireflectance plots. Figures
40 and 41 present reflectivity data from acenaphthylene carbon as
a function of orientation angle Θ. From these reflectivity data
using Eq. (99) it is possible to calculate the optical parameters
n and k. The results of these calculations are shown in Figs. 42
to 44.

Equation (93) defines the anisotropy index Ω. Figure 45 shows
the variation of Ω with HTT for two carbons. Equations (95) and (96)

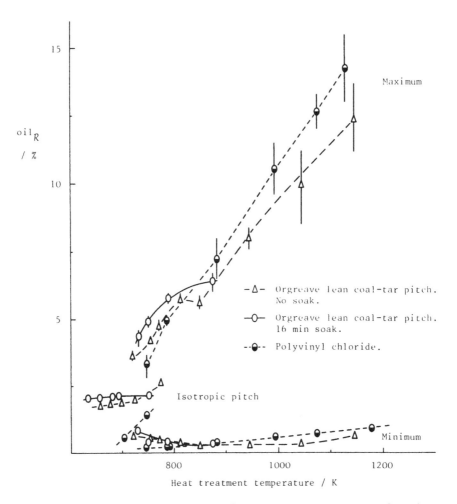

FIG. 35 Reflectivity data in oil for anisotropic carbons of various heat-treatment-temperature values.

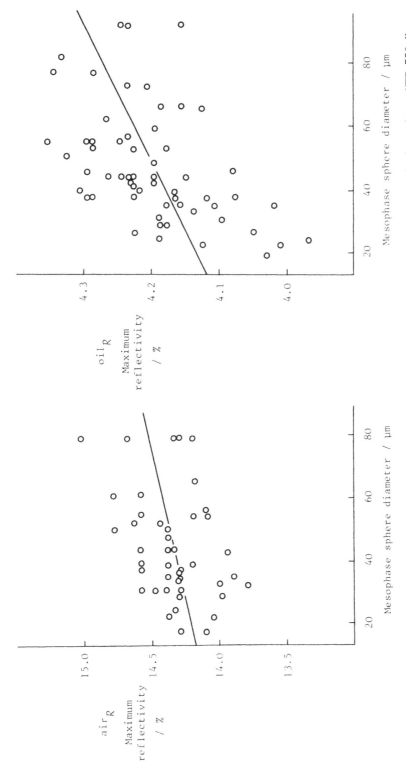

FIG. 36 Orgreave lean coal-tar pitch carbon, HTT 730 K. The variation of maximum air reflectivity with diameter of sphere of mesophase in isotropic pitch. For points with this scatter the product moment correlation coefficient r = 0.43.

FIG. 37 Orgreave lean coal-tar pitch carbon, HTT 730 K. The variation of maximum oil reflectivity with diameter of sphere of mesophase in isotropic pitch. For points with this scatter the product moment correlation coefficient r = 0.47.

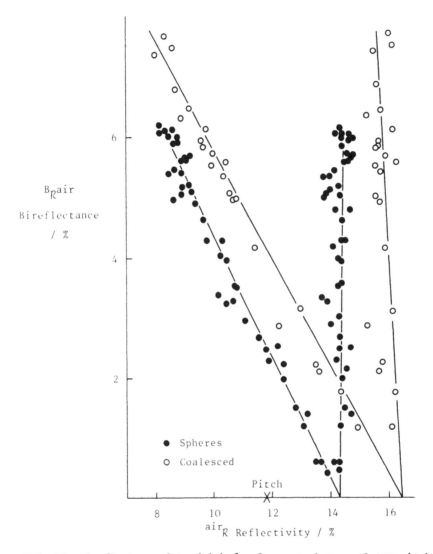

FIG. 38 Bireflectance data (air) for Orgreave lean coal-tar pitch
carbon, HTT 730 K.

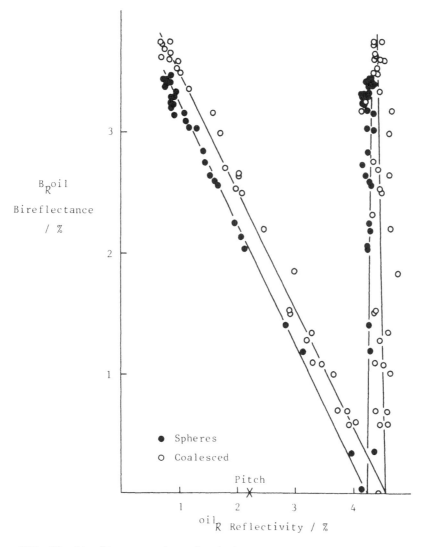

FIG. 39 Bireflectance data (oil) for Orgreave lean coal-tar pitch carbon, HTT 730 K.

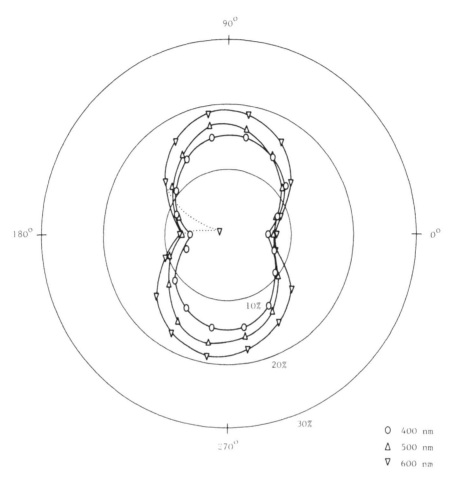

FIG. 40 Reflectivity measurements in air on the polished surface
of a carbon prepared from acenaphthylene, HTT 873 K. The value of
percentage reflectivity (^{air}R) is plotted radially for different
rotation angles (Θ). The orientation of the carbon is approximately prismatic edge ($\Phi = \pi/2$).

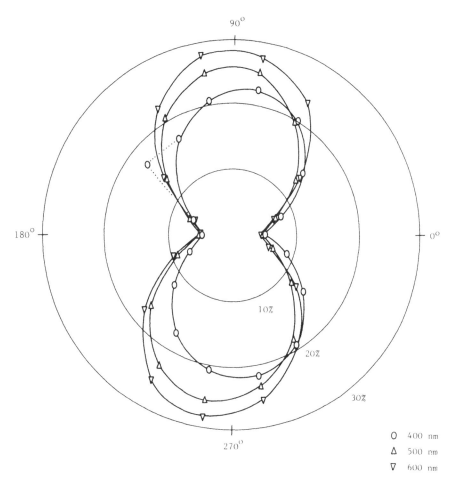

FIG. 41 Reflectivity measurements in oil on the polished surface of a carbon prepared from acenaphthylene, HTT 873 K. The value of percentage reflectivity (^{oil}R) is plotted radially for different rotation angles (Θ). The orientation of the carbon is approximately prismatic edge ($\Phi = \pi/2$).

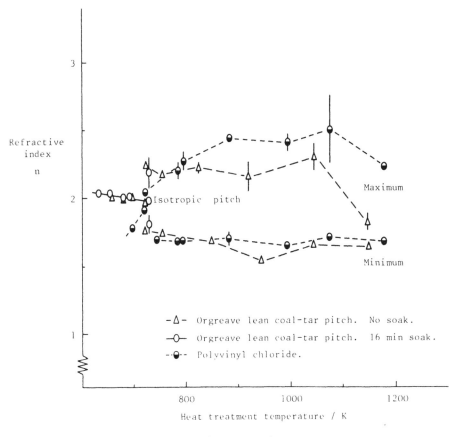

FIG. 42 Refractive index as a function of heat-treatment temperature
of various carbons.

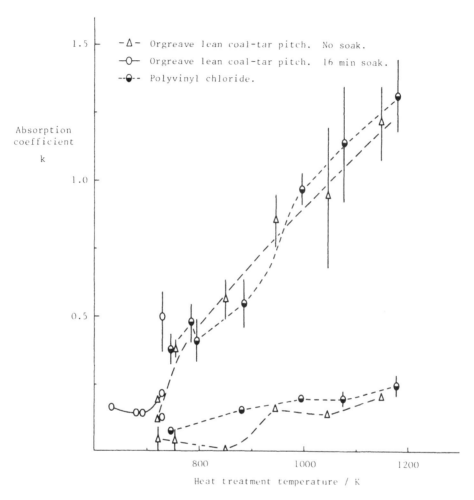

FIG. 43 Absorption coefficient as a function of heat-treatment
temperature of various carbons.

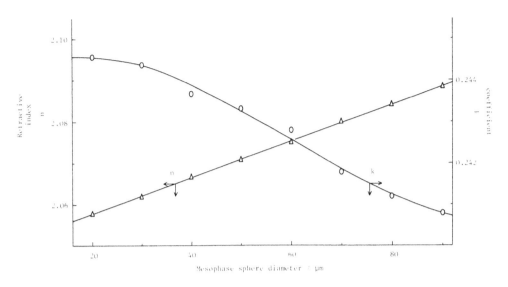

FIG. 44 Variation of n and k with diameter of spheres of mesophase;
Orgreave lean coal-tar pitch carbon, HTT 730 K.

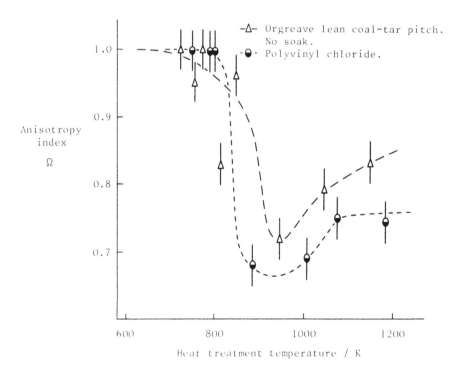

FIG. 45 Variation of anisotropy index Ω with heat treatment temperature for polyvinyl chloride carbon and Orgreave lean coal-tar pitch
carbon (no soak).

can be used to calculate the true values of reflectivity and from these the true optical parameters can be calculated. Figures 46 to 49 present these results.

Figure 40 presents reflectivity data as a function of orientation angle Θ for acenaphthylene carbon at three wavelengths. Using these data, values of n and k as functions of Θ can be calculated as shown in Figs. 50 and 51. Figure 52 shows these dispersive properties of refractive and absorptive indices as a function of wavelength.

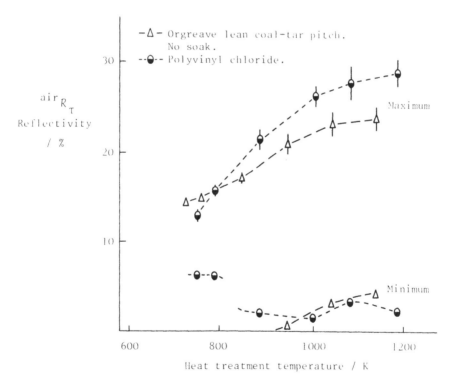

FIG. 46 Calculated true values of reflectivity (structure free) in air as a function of heat-treatment temperature for various carbons.

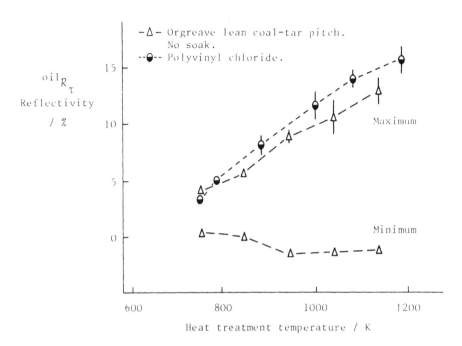

FIG. 47 Calculated true values of reflectivity (structure free) in oil as a function of heat-treatment temperature for various carbons.

D. Discussion of Results (Part 1)

*1. Orgreave Lean Coal-Tar Pitch Carbons,
     HTT 630 to 788 K*

The first growth of mesophase spheres at 681 to 730 K is not accompanied by any significant rise in air and oil reflectivites of the pitch phase (Figs. 34 and 35) possibly because of the removal of the larger molecules into the mesophase systems as they are formed in the pitch phase.

The effect of soak time on the optical texture of mesophase seems to be one of increasing the complexity of the optical texture of the mesophase. The carbon, HTT 730 K, 16-min soak period, shows a range of size of optical texture from 0.5-μm-diameter spheres through coalescing spheres of mesophase to bulk coalesced mesophase up to 50 μm in size of isochromatic domains. This contrasts markedly with the no-soak carbons from this pitch where at a similar 70 percent

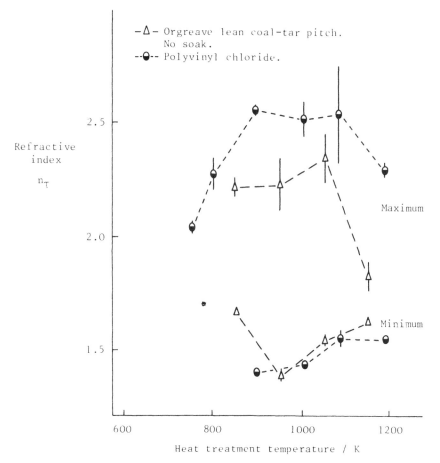

FIG. 48 Calculated true values of refractive index n_τ as a function of heat-treatment temperature for various carbons.

conversion at 756 K, the mesophase consisted of regularly arrayed spheres of uniform size (20 to 30 µm diameter).

For this pitch carbon, HTT 730 K, the calculated mean values of n and k are higher than measured for the pitch surface, which may possibly be due to the higher density of the mesophase as well as increased conjugated bonding in the molecules which constitute the mesophase (mean k calculated is 0.17, compared with the experimental value of 0.11). The fall in $^{oil}R^{min}$ from 730 K to 788 K and the

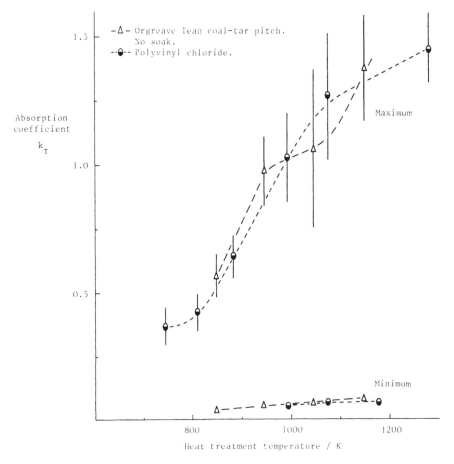

FIG. 49 Calculated true values of absorption coefficient k_τ as a function of heat-treatment temperature for various carbons.

corresponding increase in $^{oil}R^{max}$ and $^{air}R^{max}$ in Fig. 35 indicates an improvement in the alignment of the absorbing elements.

Pitch carbon HTT 730 K was selected for further investigation because of the availability of the range of size of optical texture of the mesophase exhibiting variations in reflectivities. Reflectivities for coals and cokes may be represented (Goodarzi and Murchison [53]) as a mean maximum value (e.g., mean of the top 20 percent of values), or a mean of all maximum values or a maximum maximum

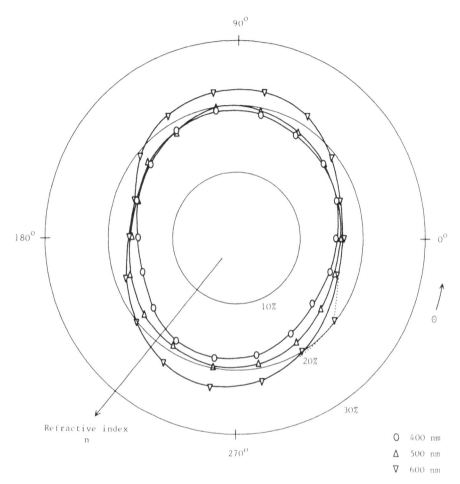

FIG. 50 Variation of refractive index as a function of orientation angle θ for the reflectivity data of Fig. 40.

value. With these pitch carbons it was considered that the variations in reflectivity were beyond experimental error and were due
to structural differences. Values of reflectivities are shown as
a bar where this is appropriate. For the carbon HTT 730 K, reflectivities of spheres and coalesced material showed that $^{oil}R^{max}$ was
higher and $^{oil}R^{min}$ was lower for coalesced mesophase than for spherical mesophase. Also, the correlation of sphere diameter with

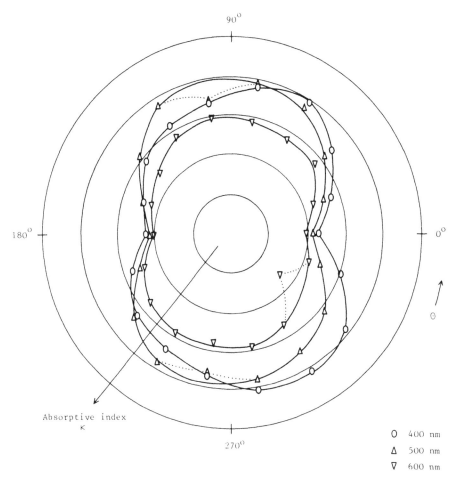

FIG. 51 Variation of absorptive index as a function of orientation
angle Θ for the reflectivity data of Fig. 40.

optical properties (Figs. 36 and 37) shows that $^{oil}R^{max}$ and $^{air}R^{max}$
reflectivities increase with increasing diameter.

The net effect of coalescence appears then to increase the
degree of ordering within the mesophase, since only this can both
increase $^{oil}R^{max}$ and decrease $^{oil}R^{min}$. The spheres and coalesced
material can be juxtapositioned and hence are of identical HTT.
The small differences between the calculated mean values of n and
k for spheres and coalesced mesophase indicate that this improvement

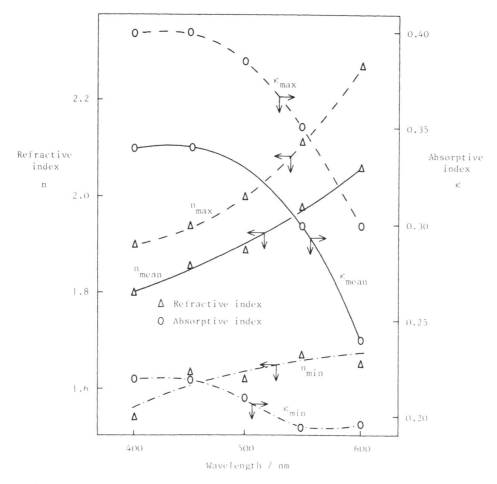

FIG. 52 Acenaphthylene carbon, HTT 873 K. The variation (dispersion) of refractive and absorptive indices with wavelength of reflected polarized light.

in alignment of the absorbing elements is not accompanied by any marked change in the properties of the absorbing elements. It is suggested that enhanced facility of mesophase to flow in the larger coalesced material may increase the degree of ordering.

2. *Orgreave Lean Coal-Tar Pitch Carbons,*
   HTT 650 to 1200 K, No Soak Period

a. *Reflectivity measurements.* This no-soak series was made to minimize observed variations of optical texture of mesophase associated

with soak time as observed in the previous series. The soak period
reduced the textural heterogeneity and gave reflectivities higher
by about 2 and 0.5 percent for air and oil, respectively (Figs. 34
and 35). The optical textures observed here are in broad agreement
with the studies of Brooks and Taylor [9] and White [11], in particular the growth of mesophase and the formation of flow-type structures
and disclinations.

No reflectivity data are available in the literature for pitch
cokes. Both the air and oil reflectivities reported here are notably
higher (compare Figs. 34 and 35 with 33) than reported for cokes from
coals of equivalent heat-treatment temperature [45,51,53,54] (with
one exception). The reflectivity values of the literature were reported without consideration of the effects of mozaic size relative
to the measuring aperture of the photometer. Since the size of the
anisotropic units varies with the initial rank of the coal, the temperature, and the heat-treatment procedure (Patrick et al. [55]),
the reported values will be influenced by the extent of supramozaic
alignment. The one exception is the data of de Vries et al. [52],
who report that whereas the oil reflectivities of cokes of HTT 1273 K
from high and low rank coals are in the range 8 to 10 percent, those
from coking coals rise up to as high as 16 percent.

It is probable that the larger anisotropic units developed by
coking coals allow measurement of a true maximum reflectivity. Lower
oil reflectivities reported by Goodarzi and Murchison [53] for coals
of equivalent ranks, heated to the same temperature range, could
result from use of larger measuring apertures or different procedures
for calculating maximum reflectivity values.

The rapid rise in R^{max} from about 756 K coincides with the temperature at which the spheroidal to coalesced mesophase transition
occurs. A detailed search of this carbon of HTT 723 K found a limited number of larger spheres and coalesced units. The oil reflectivity measurements which were made on these larger sizes of optical
texture show the same variation in R^{max}, coinciding with a rise in
bireflectance, suggests again that the major difference in reflec-

tivity between the coalesced and spheroidal mesophase is due to the
perfecting of stacking during coalescence.

The plateau observable in the maximum reflectivity curves for
both air and oil (Figs. 34 and 35) at about 850 K coincides with the
modification of the coalesced domains of optical texture into a flow-
type structure. This modification also coincides with an increase
in the variability of the measured values of maximum air and oil
reflectivity for any given carbon and a rise in $^{oil}R^{min}$. Throughout
this temperature region the C/H ratio continues to rise steadily and
the calculated mean reflectivity continues to increase relatively
smoothly. This also supports the conclusion that the inflections in
the experimental reflectivity curves are of a structural rather than
a chemical origin.

b. Refractive and absorptive indices. For a more detailed analysis,
the changes of the refractive index n, and the absorptive index k,
with HTT must be considered. The plots of the refractive index and
absorption coefficient measured at 546 nm against heat-treatment
temperature (Figs. 42 and 43) can be interpreted in terms of anomalous dispersion trends (Fig. 14) expected for an absorption band or
edge passing the 546-nm window from high energy to lower energy with
increasing heat-treatment temperature. The maximum absorption rises
relatively uniformly from very low values (e.g., 0.2) of the pitch
up to about 1.1 to 1.35 for the carbons of HTT 1148 K, while the
maximum refractive index initially constant at about 2.1, rises to
about 2.3 in carbons of HTT 1045 K, and then falls to about 1.8 in
carbons of 1145 K. The general trends shown by these data are similar to those shown by other studies (Goodarzi and Murchison [53,54]),
but the values of n and k are higher in this study and reversal in
the slope of n^{max} occurs here at higher temperatures (compare Figs.
42 and 43 with Fig. 33).

The relatively large changes in values n and k are consistent
with major changes in the electronic properties of the absorbing
elements. It is significant to note that the observed properties
for relatively low temperature pitch cokes approach those observed

for graphite at this wavelength, where $^{air}R^{max} \approx 29$ percent. It
would be erroneous to assume that this indicates that the size of
the absorbing elements in the carbon of HTT 1045 K is so large that
they do not differ significantly from that of an infinite graphite
layer. Indeed, the fall in the refractive index to 1.8 with an HTT
of 1148 K confirms that the similarities with graphite are fortuitous.
The dispersion at this HTT may be envisaged as similar to that depicted in Fig. 29b, where the visible light optical properties are
dominated by aromatic electron oscillations. This is broadly in
agreement with the electron microscopy results where carbons of HTT
1300 to 1500 K show areas of short aligned fringes, in contrast to
the larger sheets observed above this temperature range. Thus the
rise in reflectance over visible wavelengths, up to 1200 K, is due
primarily to the migration of an absorption edge to lower energies
(longer wavelengths). A similar fall in refractive index about 873 K
has been noted by Goodarzi and Murchison [53,54], both for vitrinites
and oxidized vitrinites of coals from 82.5 to 93.1 percent carbon,
although, as noted earlier, the absolute values of n and k were
lower. The maxima for the refractive indices occur, for these
vitrinite cokes, about 80 to 100 K above the initial rise in the
absorption at 673 K.

That the fall in refractive index is indeed due to anomalous
dispersion is confirmed by the dispersion data of Marshall and
Murchison [45], who showd that not only was the refractive index of
cokes of HTT 1023 K lower at all wavelengths in the visible region
than cokes of HTT 873 K from vitrinites of carbon content 83.1, 88.6,
and 93.3 percent, but that the slope of the refractive index dispersion had become much less positive in going from the 873-K to 1023-K
cokes. The point of inflection in the pitch-coke refractive indices
occurs at about 1100 K (Figs. 42 and 43), some 200 K higher than
reported for coal (e.g., Goodarzi and Murchison [53]).

c. Interpretations of structure. This difference between pitch
coke and coal coke could be due to the latter containing larger
aromatic molecules, so giving rise to absorption in the visible

region of the spectrum at lower values of HTT. This seems reasonable
since the coal-tar pitch is a low molecular weight distillate of coal.
The refractive index of the pitch cokes rises to higher values and
then falls more rapidly to lower values compared with coal cokes,
which indicates a narrower absorption band in the former.

The fall in minimum refractive index, which corresponds to a
rise in minimum absorptive index over the temperature range 850 to
945 K, could be interpreted as the development of electronic absorption and a corresponding anomalous refraction dispersion at 546 nm
for light polarized at right angles to the plane of the absorbing
elements. But it is felt more likely that the minimum refractive
index value at 945 K is probably erroneously low and the fall in n^{min}
from about 1.75 at 700 K to 1.65 at 1148 K is regular and can be
attributed to changes in ordering and density induced by heat treatment.

The significance of the changes observed in k^{max} are best discussed by reference to changes in the anisotropy index, Ω, with heat-
treatment temperature shown in Fig. 45. This figure shows that the
anisotropy index defined by Eq. (93) is near unity up to about 850 K,
when it falls sharply to a minimum of 0.72 at 945 K, from which it
rises to 0.84 at 1148 K. This dramatic fall in Ω occurs over the
same temperature range as does the breakup of the coalesced domains
into the finer flow texture and appears to be evidence of the correlation of textural disruption with inferred changes in alignment of
absorbing elements on the molecular level. The rise in Ω, after
945 K, could also be interpreted as evidence of an annealing or
slowly improving alignment.

Huttinger and Rosenblatt [56] noted the reduction in x-ray
crystallite height, L_c, over this temperature range for pitch cokes,
and this may well be the same phenomenon of reduction in order at
the absorbing element level. Cleavage also develops at about this
temperature (HTT 814 K onward) and it seems possible that stresses
set up during the molecular and textural disruptions may be related
to the development of cleavage.

d. Errors in measurements. It is important to consider the large
errors that are attendant on making measurements of absorption with
the microscope (Piller and Van Gehler [57]). From the reflectivity
values given in Figs. 34 and 35, where the typical errors are about
2.5 percent, a number of the derived absorption coefficients were
calculated to be imaginary. A similar result was noted by Ergun [1]
for absorption data derived from the minimum reflectivity of single
crystals of graphite. This seems to occur only when k might be expected to be small. Although the errors in reflectivity measurements
are quite large for small values of reflectivity, the effect of a
given error in oil and air reflectivity on k is small for small
values of k. Since k is small for maximum and minimum values of
carbons with heat-treatment temperatures of 723 K, the values of Ω
in this region must be of little significance but from the data presented here they appeared to be near unity. The values of k^{min} never
exceeded about 0.2 and hence the error in Ω must be large, but it is
significant that with one exception at an HTT of 814 K the trend
shown by Ω is quite consistent.

It may be possible to obtain more accurate values for the anisotropy index by using reflectivity values taken from bireflectance-
reflectivity plots coupled with the use of oil and air reflectivity
data taken only from a particular type of textural unit. Such an
approach would be ideally suited to a microscope with on-line com-
puting facilities available for data processing.

By applying Eqs. (95) and (96) to the experimental maximum and
minimum values of R, n, and k and using the parameters Ω, the appropriate true values R_τ, n_τ, and k_τ were calculated (Figs. 46 to 49)
for the pitch coke series (no soak). These plots show a progressive
rise with HTT. The fall in the maximum refractive index in carbons
of HTT > 1000 K remains a dominant feature and confirms the earlier
conclusion that it is an electronic feature and is not related to
the stacking modes of the absorbing elements.

e. Optical properties and density. It has been suggested (McCartney
and Ergun [42]) that the effects of density changes can be eliminated

by the use of the specific reflectivity defined as R/ρ, where ρ is
the density. No density measurements were made in this study, but
White [11] has collected data from a number of studies and has noted
that coal-tar pitch generally has a density of about 1.33 g cm^{-3} and
first-formed mesophase of about 1.40 g cm^{-3}. Correction of the pitch
reflectivities and mean reflectivities calculated from the coexisting
mesophase does reduce the discrepancies between these two figures,
suggesting that much of the apparently higher mean reflectivities
observed on mesophase are due to the rise in density.

3. Polyvinyl Chloride Carbons

The values and trends of the optical properties of the polyvinyl
chloride (PVC) carbons are essentially similar to those of Orgreave
lean coal-tar pitch observed with the coke, series 2. Certain small
differences exist. The isotropic pitch phase first formed from PVC
at about 650 K contained two coexistant phases one light (^{air}R =
6.65 percent), one dark (^{air}R = 4.98 percent) (Figs. 34 and 35).
These two phases probably represent progressive stages in the pyrolysis process, the lighter phase having a higher reflectivity, higher
refractive index, and probably greater aromaticity.

a. Reflectivity measurements. In contrast to the Orgreave pitch,
which had probably seen temperatures up to about 650 K during manu-
facture, the PVC pyrolysate showed a marked rise in reflectivity
during the isotropic stage. This rapid change in optical properties
is to be expected since large-scale chemical changes occur during
this low temperature heat treatment stage (Marsh et al. [58]). By
746 K the carbonization products contained an isotropic phase with
spheroidal and coalesced mesophase. This heterogeneity of texture
has previously been noted where a soak period has been used. The
observed reflectivity of the isotropic pitch phase at 746 K was lower
than the calculated mean for both oil and air, although a density
correction was not possible as the density data for PVC was not available. It is noticeable that the mean trend for ^{air}R and ^{oil}R shows
no inflection where mesophase is formed. The rise in reflectivity
is continuous from the lowest temperatures measured. This suggests

that the apparent plateau observed on the pitch phase for the coal-
tar pitch pyrolysates prior to mesophase formation is caused by the
earlier maximum temperature of the pitch. Goodarzi and Murchison
[53] found similar behavior in the reflectivity of carbonized coals
of different ranks.

The rapid rise of maximum air reflectivity with HTT from 743 K
to about 800 K continues the upward trend. The rate of increase
falls off from about 800 K to 1075 K and then rises slightly in the
next 500 K to give reflectivities between 28 and 31 percent at 1573 K.
Indeed, between 1075 and 1178 K the calculated mean reflectivity falls
slightly and according to other unpublished data (G. Gavari, personal
communication, 1974), a fall in maximum air reflectivity continues to
at least 2500 K. At lower heat-treatment temperatures a similar
trend is followed by the oil reflectivities.

It is significant that the values of both the maximum air and
oil reflectivities were higher than those for the coal-tar pitch
cokes at the same temperature and are greater than those of graphite
($^{air}R \approx 29$ percent). This is no criterion for stating that the two-
dimensional planar structures developed in the PVC have reached the
same level of perfection as that in graphite.

The minimum air and oil reflectivities show similar trends to
those observed for the coal-tar pitch cokes. The air minimum reflectivity falls slowly to 833 K and then rises steadily to 1573 K while
the oil minimum reflectivity falls to 786 K and then rises slowly to
1178 K.

b. Refractive and absorptive indices. The possible significance of
these changes can best be understood by reference to the variation of
the derived parameters of n and k with heat treatment. The changes
of refractive index with heat treatment for the PVC carbons (Figs.
42 and 43) show two maxima. The first occurs at about 875 K after
a very steep rise from about 2.0 at 725 K to about 2.4 at 875 K.
From the maximum at 875 K the lowest n^{max} values fall to about 2.2
at 1178 K, while the highest n^{max} values rise to give a second maximum at 1075 K. From here the value of the maximum refractive index

falls to about 2.25 at 1178 K. This second maximum at 1075 K corresponds to that observed in the coal-tar pitch cokes at the same temperature. That this second maximum is observed for the calculated
values of the mean refractive index suggests that it has electronic
significance. The origin of the first maximum is not clear, but
since it occurs at a relatively low temperature, it may be associated
with changes during loss of the chloride from the pyrolysate. It
does not correspond with any large-scale changes in maximum absorption. Values of minimum refractive index for the PVC carbons show
only a small fall in the region 900 to 1000 K, in contrast to the
larger drop observed over the same region for the coal-tar pitch
cokes.

The maximum absorption coefficient remains relatively constant
up to about 850 K, but as discussed for the coal-tar pitch cokes
(with soak), the large variation in size of optical texture found
in carbons of this temperature range when using a half-hour soak
period must introduce considerable error, particularly in the values
of k. From about 850 K the values of k^{max} increase smoothly with
temperature, giving values of 1.45 by 1178 K. While the maxima are
increasing, the minima are also rising from 0.07 at 746 K to about
0.25 at 1175 K. This rise appears to be due to enhanced perfection
of stacking of absorbing elements since the values of the anisotropy
index, Ω (Fig. 45) fall rapidly from near unity at 800 K to about
0.65 at 900 K. As for the pitch cokes, this fall in the mesophase
anisotropy coincides with the development of extensive flow structure (795 to 883 K) and is followed by the development of cleavage
fractures between 883 and 994 K.

Using the anisotropy index Ω, the optical properties of the
perfectly aligned absorbing elements were calculated (Figs. 46 and
49). These plots show a more pronounced maximum refractive index
at about 900 K, casting doubt on the maximum at 1073 K as a real
electronic feature. It may mean that the anomalous dispersion of
n starts at a lower temperature (i.e., 900 K) for PVC carbons compared with 1100 K for the pitch cokes.

4. Acenaphthylene Carbons

Air reflectivity measurements were made on a series of carbons prepared from acenaphthylene under pressures of 10,000 psi (65 MPa). The changes of maximum air reflectivity with temperature (Figs. 34 and 35) show similarities with those of the coal-tar pitch and PVC cokes, $^{air}R^{max}$ rising until 890 K thereafter leveling off at 950 K, the upper temperature limit of the high pressure carbonization apparatus. In the first solid product that could be polished, produced at 850 K, coexisting pitch, spheres, and coalesced materials were seen. In common with the coal-tar pitch and PVC cokes the coalesced materials showed a higher maximum and lower minimum reflectivity (i.e., a higher bireflectance) than the isolated spheres. This again confirms that coalescence brings about some improvement in ordering of the absorbing elements. Values of air reflectivities initially are lower than for coal-tar pitch and PVC, rising only to 17 percent by 900 K. The reflectivity values of the pitch phase also show a small rise from 850 K to 880 K.

The limitation of a 950 K maximum temperature of operation of the high pressure carbonization apparatus meant that this acenaphthylene series may have only reached the first plateau in the rise of reflectivity with temperature by 950 K; this plateau has been noted for the coal-tar pitch cokes. Previously, this plateau has been interpreted as a function of increased number of absorbing elements brought about by the disturbances associated with the formation of the flow-type mesophase.

Over this temperature range it is predicted above that the growth in diameter of absorbing element causes the migration of an absorption edge through the visible region. To check this, the dispersions of the refractive and absorptive indices were measured for the 873-K acenaphthylene carbon from 400 to 600 nm (Fig. 52). Reflectivity measurements were made on a single approximately prismatic section of mesophase in both air and oil. Figure 52 shows that the maximum absorption fell from about 0.4 at 400 nm to 0.3 at 600 nm, while the refractive index rose from 1.9 to 2.27 over the same

spectral range. This pattern of rise of absorption and fall in
refraction is that expected for a migrating absorption edge. Similar behavior is shown for coked vitrains in Fig. 32. It appears
that, although the refractive index of the coal-tar pitch coke fell
above 1100 K, dispersion with the acenaphthylene carbon, HTT 873 K,
would predict a falling refractive index in the region 900 K. Either
the absorption edge passes through the visible region at lower values
of HTT for the acenaphthylene or other factors, such as density
changes, have had a considerable effect.

The trends for both the minimum absorptive and refractive
indices follow those for the maximum values. This, together with
the high value for the minimum absorptive index, suggests that the
(single) section upon which the measurements were made was not truly
prismatic but some more intermediate section. In such a section the
values of the observed minimum will be influenced considerably by
the maximum values.

E. Materials Used (Part 2) and Results

1. *Ashland A200 Petroleum Pitch*

    (a) This series of carbons were prepared by heating the pre-
cursor in open boats in a nitrogen atmosphere to HTT 1100 K
at 5 K min^{-1} (no soak time). Samples of this 1100-K carbon
were further heat treated under argon in high temperature
furnaces.

    (b) This series of carbons were prepared by heating the pre-
cursor in a glass pot under nitrogen to HTT 800 K at about
50 K min^{-1} with 5 h of soak time. This 800-K green coke
was further heat treated to HTT 1273 K under nitrogen at
5 K min^{-1} (30 min of soak time) in a horizontal tube furnace
and above HTT 1273 K under argon in a graphite resistance
furnace (30 min of soak time).

    (c) This series of carbons were prepared by heating the pre-
cursor in a gold tube at a pressure of 130 MPa to an HTT of
873 K at 5 K min^{-1} and a soak time of 30 min. Higher HTT
samples could be prepared from this carbon by heating under

nitrogen in a horizontal furnace at 5 K min^{-1} and a 30-min
soak time. For further details, see Ref. 59.

2. Gilsonite Pitch

A series of carbons were prepared as in case (a) above.

3. Coal Extract

1. A series of carbons were prepared as in case (a) above
   using a solvent refined coal extract D112 prepared by the
   National Coal Board (NCB).
2. A series of carbons were prepared as in case (b) above
   using D112.
3. A series of carbons were prepared as in case (c) above
   using an extract from Linby Coal (Rank 901/2) prepared by
   the NCB.
4. A carbon was prepared as in case (c) above using an extract
   from Annesley Coal (Rank 701) prepared by the NCB. The
   resultant carbons were prepared and examined as described
   in Sec. VI.B.

F. Discussion of Results (Part 2)

Experimental results shown in Figs. 53 and 54 and Sec. VI.B established that with increasing HTT (to about 1200 K) the reflectivity
of carbon surfaces continuously increases. This section examines
carbons of higher HTT, where it may be anticipated that the increase
in reflectivity with HTT would continue, eventually reaching a plateau of values corresponding to the reflectivity of graphite. This
expectancy is based on x-ray diffraction studies or the direct imag-
ing by electron microscopy of the lamellar molecular constituents of
carbon. These indicate that the structure of carbons becomes more
extensively ordered or graphitic at higher temperatures.

1. Carbons from A200 Petroleum Pitch

For the series of carbons (a), Fig. 53 indicates that the reflectivity levels off about 1500 K, but still rises slightly to about
32 percent by HTT 2400 K. This value of reflectivity is much higher
than ^{air}R for graphite or for the pitch carbons of Sec. V.A, but is

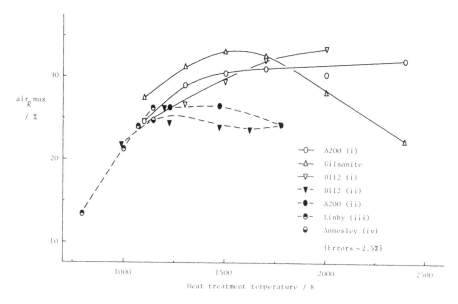

FIG. 53 Reflectivity data in air as a function of heat-treatment temperature for various carbons.

of similar magnitude to the results from PVC carbons of Sec. V.A. For carbons (b) the reflectivities are inexplicably less. Figure 54 shows the variation of ^{air}R with orientation angle Θ, the results being similar to those of Fig. 40 for acanaphthylene carbon, but here there is a larger value of anisotropy ratio (R^{max}/R^{min}) and the similarity with Fig. 8 is striking.

Table 4 shows maximum and minimum reflectivities measured on an A200 carbon HTT 1100 K [series (a)] at 10 positions. Each region, as judged by reflection interference colors, was a prismatic edge, and the theory of Sec. III.A predictsthat R^{max} and r (the reduced reflectivity) should be constants. As can be seen, there is considerable variation in both parameters, 22.0 to 28.4 percent for R^{max} and 2.034 to 4.052 for r.

This discrepancy is due either to incorrect assumptions of the theory, or to an experimental artifact—polishing damage to the surface. The theory of Sec. III.A assumed uniaxial symmetry; if this assumption is incorrect and biaxial symmetry is present in carbons,

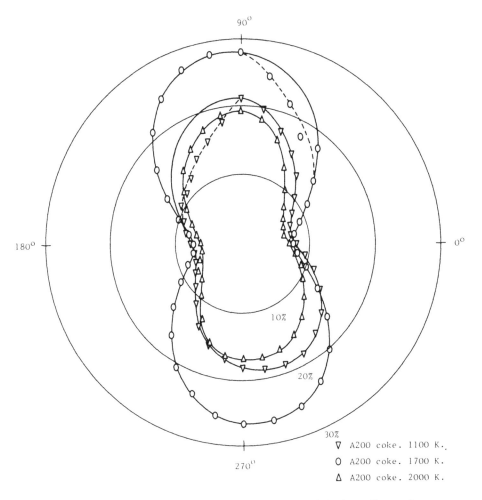

FIG. 54 Reflectivity measurements on the polished surface of a
carbon prepared from A200 petroleum pitch. The value of percentage
reflectivity (^{air}R) is plotted radially for different rotation
angles (Θ). The orientations of the carbons are approximately
prismatic edges ($\Phi = \pi/2$).

this could explain the discrepancy. It has been reported in the literature [8] that in some regions of carbons, especially flow regions,
it is possible to detect nonuniaxial symmetry by electron microscopy.
It is suggested that the theory is reduced to biaxial. However, it
is believed that any biaxial nature of the material will be of little
importance compared to the predominant uniaxial symmetry.

TABLE 4   Maximum and Minimum Reflectivities of Regions
of Coke from A200 Petroleum Pitch, HTT = 1100 K

Region number	R^{max} (%)	R^{min} (%)	$r = R^{max}/R^{min}$
1	26.0	7.8	3.333
2	28.0	7.3	3.836
3	25.5	9.2	2.772
4	23.5	5.8	4.052
5	22.0	6.5	3.385
6	24.0	11.8	2.034
7	28.4	7.8	3.641
8	25.2	7.4	3.405
9	22.0	6.7	3.284
10	25.6	7.6	3.368

If the effect of polishing on a prismatic edge and a basal
plane is considered, then intuitively the damage to a basal plane
is less than to a prismatic edge where the graphitic planes are
"bent over." This polishing damage to prismatic edges can explain
the higher reflectivities found on basal planes than on prismatic
edges, and the variation in R^{max} and r values.

If this hypothesis of polishing damage is correct, the calculation of parallel and perpendicular optical parameters is very difficult. This is because the prismatic edge ($\phi = \pi/2$) which can be
used to give the parallel *and* perpendicular parameters is unrepre-
sentative of the material because of polishing damage. The basal
plane ($\phi = 0$) is representative of the material because of little
polishing damage; however, it cannot be used to calculate the perpendicular parameters. A possible way of avoiding this problem in
carbons susceptible to polishing damage is described below.

The analysis in Section III.A shows that for a range of orientation angles near the basal plane configuration ($\phi = 0$), the eye will
be unable to detect a color change on rotation. The region appears
purple and is therefore interpreted as isotropic. An angle of 30°
with the basal-plane configuration was suggested as the extent of

deviation necessary before a color change could be detected. These
orientations are defined as near-basal-plane configurations. It is
probable that these near-basal-plane configurations have the least
polishing damage and hence are most representative of the material.
From the theoretical equations a value can be found for the average
of the minimum reflectivity values over this range of near-basal-
plane configurations. Equation (103) expresses this theoretical
average.

$$\bar{R}^{min}[\omega] = \frac{1}{\omega} \int_0^\omega R^{min}(\Phi) \, d\Phi \tag{103}$$

In Eq. (103) the notation $\bar{R}^{min}[\omega]$ denotes an average over the range
of angles ω from the basal plane. Because R^{max} is theoretically a
constant for the material, the value of $\bar{r} = R^{max}/R^{min}[\omega]$ can be obtained. Experimentally a set of values for r, $\{r_i\}$, is obtained.
To calculate an average comparable to the theoretical average, the
harmonic average of the r values, H_r, is found by

$$H_r = \frac{m}{\sum\limits_{i=1}^{m} (1/r_i)} \tag{104}$$

Equating the two values H_r and $\bar{r}$ can in principle give the
optical parameters. Equations (54) and (55) give model expressions
for $R^{min}(\Phi)$ and hence Eq. (103) can be evaluated analytically. This
gives an equation relating ρ_1^2 and ψ. A similar equation can be derived for experiments and theory with oil as the absorption medium.
From these two equations the values of ρ_1^2 and ψ can be found, and
hence the optical parameters n and k. To date no calculations have
been performed using the method above and therefore the practica-
bility of it cannot be judged.

2. *Carbons from Gilsonite Pitch*

This series of carbons indicates that the reflectivity passes
through a maximum with increasing HTT, the maximum reflectivity of
33 percent occurring at about 1500 K. The fall of reflectivity in
the range 1500 to 2500 K is very pronounced; there is no indication

that the reflectivity will again increase to reach a graphitic maximum at higher HTT. Following Forrest et al. [60], suggestions for this maximum can be made:

1. The decrease in reflectivity after the maximum could be due to the development of microfissures in the carbon which affect the amount of light reflected from the surface. Here, the fall in reflectivity does not indicate a change in actual structure of the absorbing molecular units, but rather is due to changes at the carbon surface.

2. At a certain stage of the graphitization process, the lamellar constituent molecules become strained and deformed by cross-linkages and bond-shortening mechanisms. This may produce a zigzag molecular structure which has been recognized in the high resolution, phase-contrast electron microscope [61].

3. The position of the main absorption band (usually in the ultraviolet) dictates the shapes of the $n(\lambda)$ and $k(\lambda)$ curves in the visible region. A possible movement of this band toward the visible region would affect the optical parameters and could be responsible for the fall of reflectivity, as discussed in Secs. V.A and III.A. To investigate this hypothesis reflectivities would have to be measured at several wavelengths of light.

4. The effect is due to some experimental artifact, whereby in the polishing or some other process, the basal planes are removed or distorted preferentially.

Gilsonite carbon (HTT 2000 K) contains inherent fissures which are exposed by polishing. Because this Gilsonite carbon has small isotropic regions, the fissures could reduce the size of flat, well-polished regions below that which is necessary for accurate measurement. Hence a lower reflectivity will be obtained compared with carbons of lower HTT, which are less fissured.

Some general evidence for hypothesis 3 is the fact that reflection interference colors (when the retarder plate is used) change

significantly when the value of HTT reaches 2400 K. This change is
undoubtedly due to different values of the optical parameters. If
the change of optical parameters is large, the reflectivity will
change significantly, but a quantitative assessment is not possible.

3. Carbons from Coal Extract

Results for series 1 of D112 coal extract indicate that the reflectivity is still rising at HTT 2000 K. As with the carbons from A200,
the series of D112 carbons prepared by pot carbonization have appreciable lower reflectivities. The results for the Linby and Annesley
coal extracts (3 and 4) are for reflectivities at lower HTT (<1200 K).
The results here are very similar to those for the Orgreave lean
coal-tar pitch carbon shown in Fig. 35.

VII. SUMMARY AND CONCLUSIONS

This chapter has discussed the formation of anisotropic carbon in
terms of a lamellar nematic liquid crystal phase and the subsequent
polymeric mesophase. This explanation of formation of anisotropy in
carbons is the essential model necessary to describe the optical
properties of carbons. Regions in the imperfect twisting graphitic
sheets having the uniaxial symmetry of graphite are termed local
graphitic regions. The orientation of the optic axis (z axis) of
this local region with respect to the polished surface has been
defined, and by solving Maxwell's equations across this boundary
the reflectivity of the surface has been obtained. Although formally
simple, this equation contains $\rho(\Phi)$, which is algebraically very complicated. Therefore, a simplified model in terms of a reflection
response function has been proposed by which two simple algebraic
functions for $\rho(\Phi)$ are obtained. These model equations have been
shown to be accurate for carbons in many applications.

The generation of extinction contours by means of crossed
polars has been considered and the orientations for extinction
listed. For nonextinction the variation of reflectivity with
orientation has been calculated and from this the suggestion was

put forward that the eye will interpret all regions of $\Phi \lesssim 30°$ as if they were composed of basal planes.

The generation of reflection interference colors (using a phase retarder plate) has been considered and a full explanation given of these colors for both a half-wave and a one-wave plate. This explanation involves the dispersive properties of the plate and the carbon and shows that depending on the orientation of the carbon, regions of the optical spectrum are removed by destructive interference. Thus, for a prismatic edge ($\Theta = \pi/2$), values of $\Theta = 0, \pi/2$ lead to the green region of the spectrum being removed, yielding a purple color. For $\Theta = \pm\pi/4$ the red and blue regions, respectively, are removed, yielding blue and yellow colors, respectively. The dependence of optical properties on the ordering of the basic units (absorption elements) has been considered, which led to the definition of the anisotropy index (Ω). Also, the contributions from aromatic and graphitic parts of the absorption elements have been used to explain the variation of reflectivity with HTT in terms of the movement of an absorption band toward the visible region of the spectrum.

Various averages have been defined and evaluated. The calculation of the three-dimensional average in terms of the model equations is especially important.

A review has been made of optical data in the literature for cokes prepared from vitrinite and these data have been interpreted in terms of the optical theory above. New experimental results on a variety of anisotropic carbons have been presented and interpreted in terms of the optical theory. The majority of experimental and theoretical points agree closely. One worrying discrepancy, however, is the variation of R^{max} at regions of the carbon having different orientation angle Φ. Theory predicts this to be a constant. Several possible explanations for this variation have been presented, but it seems most likely that polishing damage to the surface is responsible.

These experimental studies expand the range of data enormously, but even so there still have been only two investigations of the optical properties of anisotropic carbon as a function of wavelength, and this is obviously a region for much future study.

Optical studies of anisotropic carbon are one method for studying the structure of the carbon. This method is especially useful
because of its relative experimental ease and because structures at
a larger scale inaccessible to x-ray diffraction and electron microscopy can be studied. It suffers, however, from having a less complete fundamental foundation. The optical properties of diamond and
graphite can readily be compared with the electronic band structure
of these crystalline materials. Anisotropic carbon does not have a
continuous crystalline structure, and this lack of translational
symmetry makes the solid state theory impossibly complicated.

APPENDIXES

In the following appendixes various mathematical derivations are
carried out fully.

A.1 Different Sets of Vectors

For the ordinary wave $\hat{n}\hat{s} = \pm 1$ and since by definition $\hat{n} \cdot \hat{s} = 1$, the
two vectors are collinear. From Eqs. (22) and (33), Eq. (A1) is
found.

$$\hat{s} \times H = -E \tag{A1}$$

Equations (20), (22), and (A1) show that the vectors $\hat{n}$, $\hat{s}$, E, and $\hat{D}$
are coplanar with H normal to this plane. The vectors $\{H, E, \hat{s}\}$ and
$\{H, \hat{D}, \hat{n}\}$ are two sets of three mutually perpendicular vectors. Only
if $\hat{n}$ and $\hat{s}$ are collinear are the vectors E and $\hat{D}$ collinear. Thus for
the ordinary wave E and $\hat{D}$ are collinear.

A.2 Vectors in and Normal to the Principal Plane

Landau and Lifshitz show that the vectors E_o and E_e and $\hat{D}_o$ and $\hat{D}_e$
are perpendicular. Thus it is possible to determine which vectors
lie in the principal plane and which are normal to it. $\hat{n}_o$ depends
only on $\hat{\varepsilon}_\parallel$ (not on $\hat{\varepsilon}_\perp$); therefore E_e and $\hat{D}_e$ are normal to the principal plane and E_o and $\hat{D}_o$ lie in it. Since H_e, $\hat{D}_e$, and $\hat{n}$ are perpendicular, H_e is also normal to the plane. Since H_e, E_e, and $\hat{s}$
are perpendicular, $\hat{s}$ lies in the plane. Therefore, the set of vectors $\{E_e, \hat{D}_e, H_o, \hat{n}, \hat{s}\}$ lie in the principal plane and the set
$\{E_o, \hat{D}_o, H_e\}$ are normal to the principal plane.

A.3   Relationship Between $\underset{\sim}{H}$ and $\underset{\sim}{E}$ for Ordinary Wave

Because $\underset{\sim}{E}_O$ is perpendicular to $\hat{\underset{\sim}{n}}$, this derivation is easy. From Eq. (20),

$$\underset{\sim}{H}_O = \hat{\underset{\sim}{n}}_0 \times \underset{\sim}{E}_O$$

From Eq. (16),

$$\hat{\underset{\sim}{n}}_0 = \frac{c}{\omega} \hat{\underset{\sim}{k}}_0$$

$$\Rightarrow \underset{\sim}{H}_O = \frac{c}{\omega} \hat{\underset{\sim}{k}}_0 \times \underset{\sim}{E}_O = \hat{n}_0 \frac{\hat{\underset{\sim}{k}}_0 \times \underset{\sim}{E}_O}{\hat{k}_0}$$

$$\Rightarrow \underset{\sim}{H}_O = \hat{\varepsilon}_\parallel^{1/2} \frac{\hat{\underset{\sim}{k}}_0 \times \underset{\sim}{E}_O}{\hat{k}_0}$$

A.4   Relationship Between $\underset{\sim}{H}$ and $\underset{\sim}{E}$ for Extraordinary Wave

Because $\underset{\sim}{E}_e$ is perpendicular to $\hat{\underset{\sim}{s}}_e$, a different approach is necessary. From Eq. (A1),

$$\hat{\underset{\sim}{s}}_e \times \underset{\sim}{H}_e = -\underset{\sim}{E}_e$$

$$\Rightarrow \hat{\underset{\sim}{s}}_e \times (\hat{\underset{\sim}{s}}_e \times \underset{\sim}{H}_e) = -\hat{\underset{\sim}{s}}_e \times \underset{\sim}{E}_e$$

$$\Rightarrow \hat{\underset{\sim}{s}}_e \times \underset{\sim}{E}_e = (\hat{\underset{\sim}{s}}_e \cdot \hat{\underset{\sim}{s}}_e)\underset{\sim}{H}_e - (\hat{\underset{\sim}{s}}_e \cdot \underset{\sim}{H}_e)\hat{\underset{\sim}{s}}_e$$

But

$$\hat{\underset{\sim}{s}}_e \cdot \underset{\sim}{H} = 0$$

$$\Rightarrow \underset{\sim}{H}_e = \frac{1}{\hat{\underset{\sim}{s}}_e^2} \hat{\underset{\sim}{s}}_e \times \underset{\sim}{E}_e$$

$$\Rightarrow \underset{\sim}{H}_e = \frac{1}{\hat{\underset{\sim}{s}}_e} \frac{\hat{\underset{\sim}{s}}_e \times \underset{\sim}{E}_e}{\hat{\underset{\sim}{s}}_e}$$

A.5   Reflection of Ordinary Wave

This derivation makes use of the boundary conditions across the surface.

$$H_t(1) = H_t(2) \tag{A2}$$

$$H_n(1) = H_n(2) \tag{A3}$$

$$E_t(1) = E_t(2) \tag{A4}$$

$$\{D_n(1) = D_n(2), \qquad j_n(1) = j_n(2)\} \Rightarrow \hat{D}_n(1) = \hat{D}_n(2) \tag{A5}$$

In Eqs. (A2) to (A5), (1) and (2) specify the medium, and the suffices n and t specify normal and tangential. The sign conventions and directions of fields are shown in Fig. 55.

Considering the continuity of the tangential fields across the boundary [i.e., Eqs. (A2) and (A4)], we have

$$E_{iv} - E_{rv} = E_{tv} \tag{A6}$$

$$H_{iu} + H_{ru} = H_{tu} \tag{A7}$$

Using Eq. (37) to transform the H fields in Eq. (A7) into E fields, we have

$$E_{iv} + E_{rv} = \hat{\varepsilon}_\parallel^{1/2} E_{tv} \tag{A8}$$

Eliminating E_{tv} from Eqs. (A6) and (A8) leads to the required reflection coefficient:

$$E_{iv} + E_{rv} = \hat{\varepsilon}_\parallel^{1/2} (E_{iv} - E_{rv})$$

$$\Rightarrow E_{rv}(\hat{\varepsilon}_\parallel^{1/2} + 1) = E_{iv}(\hat{\varepsilon}_\parallel^{1/2} - 1)$$

$$\Rightarrow E_{rv} = \frac{\hat{\varepsilon}_\parallel^{1/2} - 1}{\hat{\varepsilon}_\parallel^{1/2} + 1} E_{iv} \tag{A9}$$

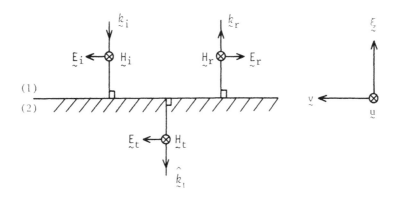

FIG. 55 Reflection of ordinary wave, directions of fields, and sign conventions.

Equation (A9) can be rewritten in the form of Eq. (25):

$$E_{rv} = \rho_o e^{i\theta_o} E_{iv} \tag{A10}$$

Equation (27) is used to express $\hat{n}_\parallel$ in terms of the real quantities:

$$\hat{n}_\parallel = \hat{\varepsilon}_\parallel^{1/2} = n_\parallel + ik_\parallel \tag{A11}$$

Equations (A9), (A10), and (A11) can then be used to give final expressions for ρ_o and θ_o in terms of $n_\parallel$ and $k_\parallel$:

$$\rho_o e^{i\theta_o} = \frac{n_\parallel + ik_\parallel - 1}{n_\parallel + ik_\parallel + 1}$$

$$\Longrightarrow \rho_o^2 = \frac{(n_\parallel - 1)^2 + k_\parallel^2}{(n_\parallel + 1)^2 + k_\parallel^2}$$

$$\tan \theta_o = \frac{2k_\parallel}{n_\parallel^2 + k_\parallel^2 - 1}$$

A.6 Reflection of Extraordinary Wave

Figure 56 shows the definition of angles θ'' and θ''' and the sign conventions for the fields. Using the tangential boundary conditions, Eqs. (A2) and (A4), we obtain

$$E_{iu} - E_{ru} = E_t \cos \hat{\theta}'' \tag{A12}$$

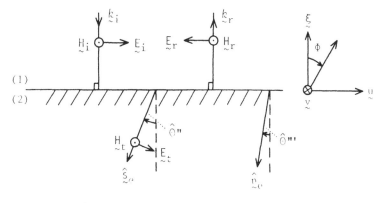

FIG. 56 Reflection of extraordinary wave, directions of fields, and sign conventions.

$$H_{iv} + H_{rv} = H_{tv} \tag{A13}$$

Using Eq. (38) to transform the H fields in Eq. (A13) into E fields yields

$$E_{iu} + E_{ru} = \frac{1}{\hat{s}} E_t \tag{A14}$$

Equation (A5) can be used to find the value of $\hat{\theta}'''$:

$$\hat{D}_n(1) = 0 \Longrightarrow \hat{D}_n(2) = 0$$

Because $\hat{\underset{\sim}{D}}$ is perpendicular to $\hat{\underset{\sim}{n}}_e$, the wave vector propagates normal to the surface, and consequently

$$\hat{\theta}''' = 0 \tag{A15}$$

The relationship between $\hat{\theta}$, $\hat{\theta}'$, and Φ can be found from Fig. 56.

$$\hat{\theta}' = \Phi - \hat{\theta}'' \tag{A16}$$

$$\hat{\theta} = \Phi - \hat{\theta}''' \tag{A17}$$

From Eqs. (A15) and (A17),

$$\hat{\theta} = \Phi \tag{A18}$$

The relationship between $\hat{\theta}$ and $\hat{\theta}'$ is given by Landau and Lifshitz:

$$\tan \hat{\theta}' = \frac{\hat{\epsilon}_{\parallel}}{\hat{\epsilon}_{\perp}} \tan \hat{\theta} \tag{A19}$$

Eliminating E_t from Eqs. (A12) and (A14) gives us

$$E_{iu} + E_{ru} = \frac{1}{\hat{s}_e \cos \theta''} (E_{iu} - E_{ru})$$

$$\Longrightarrow E_{ru} = \frac{\hat{N} - 1}{\hat{N} + 1} E_{ru} \qquad \text{where} \qquad \hat{N} = (\hat{s}_e \cos \hat{\theta}'')^{-1}$$

From Eq. (36),

$$\hat{s}_e = (\hat{\epsilon}_{\parallel} \cos^2 \hat{\theta}' + \hat{\epsilon}_{\perp} \sin^2 \hat{\theta}')^{-1/2}$$

From Eq. (A16),

$$\cos \hat{\theta}'' = \cos(\Phi - \hat{\theta}')$$

$$\Rightarrow \hat{N}^{-1} = \frac{\cos(\Phi - \hat{\theta}')}{(\hat{\epsilon}_\parallel \cos^2\hat{\theta}' + \hat{\epsilon}_\perp \sin^2\hat{\theta}')^{1/2}}$$

$$\Rightarrow \hat{N} = \frac{(\hat{\epsilon}_\parallel + \hat{\epsilon}_\perp \tan^2\hat{\theta}')^{1/2}}{\cos\Phi + \sin\Phi \tan\hat{\theta}'}$$

Using Eqs. (A18) and (A19), we obtain

$$\tan\hat{\theta}' = \frac{\hat{\epsilon}_\parallel}{\hat{\epsilon}_\perp} \tan\Phi$$

$$\Rightarrow \hat{N} = \frac{(\hat{\epsilon}_\parallel + \hat{\epsilon}_\perp \tan^2\hat{\theta}')^{1/2}}{\cos\Phi + (\hat{\epsilon}_\parallel/\hat{\epsilon}_\perp) \sin\Phi \tan\Phi}$$

$$= \hat{\epsilon}_\parallel^{1/2} \frac{[\cos^2\Phi + (\hat{\epsilon}_\parallel/\hat{\epsilon}_\perp) \sin^2\Phi]^{1/2}}{\cos^2\Phi + (\hat{\epsilon}_\parallel/\hat{\epsilon}_\perp) \sin^2\Phi}$$

$$\Rightarrow \hat{N} = \left(\frac{\cos^2\Phi}{\hat{\epsilon}_\parallel} + \frac{\sin^2\Phi}{\hat{\epsilon}_\perp}\right)^{-1/2}$$

A.7 Products of Complex Quantities

The complex notation used in this theory is not able to handle directly nonlinear expressions such as $\underset{\sim}{E} \cdot \underset{\sim}{H}$. Before forming products such as these, the fields must be converted to their real form and the products formed from these. However, because the intensity is a product of a quantity with itself, a simplification is possible [62].

$$\underset{\sim}{E}_{physical} = Re(\underset{\sim}{E}e^{i\omega t})$$

The vector $\underset{\sim}{E}$ contains the space dependence: that is,

$$\underset{\sim}{E} \equiv \underset{\sim}{E}(\underset{\sim}{r})$$

$$\Rightarrow \underset{\sim}{E}_{physical} = \frac{1}{2}(\underset{\sim}{E}e^{-i\omega t} + \underset{\sim}{E}^* e^{i\omega t})$$

$$\Rightarrow \underset{\sim}{E}_{physical} \cdot \underset{\sim}{E}_{physical} = \frac{1}{4}(\underset{\sim}{E} \cdot \underset{\sim}{E}e^{-2i\omega t} + \underset{\sim}{E}^* \cdot \underset{\sim}{E}^* e^{2i\omega t} + 2\underset{\sim}{E} \cdot \underset{\sim}{E}^*)$$

$$\Rightarrow \underset{\sim}{E}_{physical} \cdot \underset{\sim}{E}_{physical} = \frac{1}{2} Re(\underset{\sim}{E} \cdot \underset{\sim}{E}e^{-2i\omega t}) + \frac{1}{2}\underset{\sim}{E} \cdot \underset{\sim}{E}^* \qquad (A20)$$

Because all intensities actually measured are time averages, the
first term in Eq. (A20) vanishes.

$$\Rightarrow \ I \ (\equiv \overline{I}) \ = \frac{1}{2}\underset{\sim}{E} \cdot \underset{\sim}{E}^* \tag{A21}$$

A.8 Derivation of Reflectivity Equation

At the polished surface the phase of the wave is ϕ. For the incident
light, $\underset{\sim}{E}_i = Ee^{i\phi}\underset{\sim}{\eta}$. Converting from the (η,ζ) system to the (u,v)
system, we have

$$\underset{\sim}{E}_i = Ee^{i\phi}(\cos \Theta \underset{\sim}{u} - \sin \Theta \underset{\sim}{v})$$

The electric vector of the reflected light is found by use of
the reflection coefficients for the ordinary and extraordinary waves:

$$\underset{\sim}{E}_r = Ee^{i\phi}(\cos \Theta \hat{\rho}_e\underset{\sim}{u} - \sin \Theta \hat{\rho}_o\underset{\sim}{v}) \tag{A22}$$

Converting Eq. (A22) back into the (η,ζ) system gives us

$$\underset{\sim}{E}_r = Ee^{i\phi}(\underset{\sim}{\eta}\{\cos^2\Theta\rho(\Phi)e^{i\theta} + \sin^2\Theta\rho_\parallel e^{i\theta_\parallel}\}$$

$$+ \ \underset{\sim}{\zeta}\{\cos \Theta \sin \Theta\rho(\Phi)e^{i\theta} - \cos \Theta \sin \Theta\rho_\parallel e^{i\theta_\parallel}\}) \tag{A23}$$

To find the reflectivity, the intensity of the reflected light must
be calculated.

In Eq. (A23) the harmonic time term is contained in the factor
$e^{i\phi}$; the remainder of Eq. (A23) corresponds to the term $\underset{\sim}{E}$ used in
Eqs. (A20) and (A21).

$$I_r = \frac{1}{2}\underset{\sim}{E}_r \cdot \underset{\sim}{E}_r^*$$

But

$$I_i = \frac{1}{2}E^2$$

$$\Rightarrow R = \frac{\underset{\sim}{E}_r \cdot \underset{\sim}{E}_r^*}{E^2} \tag{A24}$$

Equation (A24) is the expression for reflectivity, and by using Eq. (A23) the reflectivity can be shown explicitly. Because the two vectors $\underset{\sim}{\eta}$ and $\underset{\sim}{\zeta}$ are orthogonal unit vectors, the following rules apply: $\underset{\sim}{\eta} \cdot \underset{\sim}{\eta} = \underset{\sim}{\zeta} \cdot \underset{\sim}{\zeta} = 1$ and $\underset{\sim}{\eta} \cdot \underset{\sim}{\zeta} = 0$.

$$\Rightarrow R = \cos^4\Theta \rho^2(\Phi) + \sin^4\Theta \rho_\|^2 + 2 \sin^2\Theta \cos^2\Theta \rho_\| \rho(\Phi) \cos(\theta - \theta_\|)$$
$$+ \sin^2\Theta \cos^2\Theta(\rho^2(\Phi) + \rho_\|^2 - 2\rho_\| \rho(\Phi) \cos(\theta - \theta_\|))$$

The interference term $\cos(\theta - \theta_\|)$ cancels out, and after simplification:

$$R = \rho_\|^2 \sin^2\Theta + \rho^2(\Phi) \cos^2\Theta$$

A.9 Derivation of Minimum in Intensity Spectrum

$$\Delta I_\| = \frac{2}{I_0 \rho(\Phi)\rho_\|} (I_{\|+} - I_{\|-})$$

From Table 2,

$$\Delta I_\| = \cos(\chi - \psi) - \cos(\chi + \psi)$$
$$\Rightarrow \Delta I_\| = 2 \sin \chi \sin \psi(\Phi)$$

Since $\chi(\lambda_0) = -\pi \Rightarrow \Delta I_\|(\lambda_0) = 0$. Consider

$$\lambda = \lambda_0 + \delta\lambda, \quad \delta\lambda > 0$$
$$\Rightarrow \Delta I (\lambda_0 + \delta\lambda) = 2 \sin \chi(\lambda_0 + \delta\lambda) \sin \psi(\lambda_0 + \delta\lambda)$$

Figure 16 shows that $\chi(\lambda_0 + \delta\lambda) = -\pi + \alpha$, $\alpha > 0$. Figure 19 shows that $\psi(\lambda_0 + \delta\lambda) = \psi_0 + \beta$, $\beta < 0$, where

$$\psi_0 = \psi(\lambda_0)$$
$$\Rightarrow \Delta I_\|(\lambda_0 + \delta\lambda) = 2 \sin(-\pi + \alpha) \sin(\psi_0 + \beta)$$
$$= -2 \sin \alpha \sin(\psi_0 + \beta)$$

Figure 19 shows that $0 \leq \psi_0 + \beta \leq \pi/2 \Rightarrow \sin(\psi_0 + \beta) > 0$. Figure 16 shows that α is small $\Rightarrow \sin \alpha > 0 \Rightarrow \Delta I_\|(\lambda_0 + \delta\lambda) > 0$. A similar argument shows that $\Delta I_\|(\lambda_0 - \delta\lambda) > 0$.

A.10 Relationships Between Several Optical Parameters

In calculations it is necessary to be able to express the reflection coefficient $\hat{\rho}$ in terms of two different sets of variables $\{\rho,\theta\}$ and $\{n,k\}$. From Eq. 43,

$$\hat{\rho} = \frac{\hat{n} - 1}{\hat{n} + 1} \tag{A25}$$

where $\hat{\rho} = \rho e^{i\theta}$ and $\hat{n} = n + ik$. Inverting Eq. (A25) yields

$$\hat{n} = \frac{1 + \hat{\rho}}{1 - \hat{\rho}}$$

$$\Rightarrow n + ik = \frac{1 + \rho \cos \theta + i\rho \sin \theta}{1 - \rho \cos \theta - i\rho \sin \theta}$$

$$\Rightarrow n = \frac{1 - \rho^2}{1 - 2\rho \cos \theta + \rho^2} \tag{A26}$$

and

$$k = \frac{2\rho \sin \theta}{1 - 2\rho \cos \theta + \rho^2} \tag{A27}$$

Rewriting Eqs. (40) and (41), we have

$$\rho^2 = \frac{(n - 1)^2 + k^2}{(n + 1)^2 + k^2} \tag{A28}$$

$$\tan \theta = \frac{2k}{n^2 + k^2 - 1} \tag{A29}$$

The four equations (A26) to (A29) give the relationship between the two sets of variables. Equation (A29) alone is ambiguous because θ is only defined to within π. There is no ambiguity in Eqs. (A26) and (A27) since sin and cos have periods of 2π. The use of all four equations shows that $\theta = \tan^{-1}(x) + \pi$, which is self-consistant with the sign conventions defined in Figs. 55 and 56.

A.11 Results for a General Mechanism

In all the results derived so far [e.g., Eqs. (A28) and (A29)] the refractive index of the surrounding medium is assumed to be 1.0. This is not true generally. If Appendixes A.5 and A.6 are recalculated using the general refractive index (n_i), it is found that Eq. (43) can be rewritten as

$$\hat{\rho} = \frac{\hat{N} - n_i}{\hat{N} + n_i} \tag{A30}$$

and consequently Eqs. (40) and (41) become

$$\rho_o^2 = \frac{(n_\parallel - n_i)^2 + k_\parallel^2}{(n_\parallel + n_i)^2 + k_\parallel^2} \tag{A31}$$

$$\tan \Theta_o = \frac{2k_\parallel n_i}{n_\parallel^2 + k_\parallel^2 - n_i^2} \tag{A32}$$

LIST OF SYMBOLS

Unit vectors are defined by the symbol ^ (e.g., $\hat{\underset{\sim}{n}}$ in Sec. II); complex quantities are defined by the same symbol in the remainder of the chapter.

Vectors are denoted by ~ (e.g., $\underset{\sim}{a}$) and tensors by _ (e.g., $\underline{b}$). $\underset{\sim}{\nabla}$ denotes the differential vector operator (del); in the (x,y,z) system it is $\partial(\underset{\sim}{x})/\partial x + \partial(\underset{\sim}{y})/\partial y + \partial(\underset{\sim}{z})/\partial z$.

The suffices $\parallel$, $\perp$, i, r, t, a, p, o, e, ob, and τ have the following special meanings:

$\parallel$, $\perp$ denote quantities measured parallel or perpendicular to the graphite planes.

i, r, t denote incident, reflected, and transmitted waves.

a, p denote analyzer and retarder plate.

o, e denote ordinary and extraordinary waves.

ob, τ denote observed and true optical parameters.

The superfixes max and min denote maximum and minimum parameters, while the superfixes air and oil denote reflectivities measured in air or oil.

$\hat{\underset{\sim}{a}}$ Direction of long axis for rodlike molecules

A, B   Constants defined in the text

b_R	Bireflectance ratio
$\underset{\sim}{B}$	Magnetic flux density (magnetic induction)
B_R	Bireflectance
c	Speed of light in a vacuum
d	Interlayer spacing in smectic phase
$\underset{\sim}{D}$	Electric displacemnt
e	Charge on electron
$\underset{\sim}{E}$	Electric vector
g_i, A, B, C, $\mathcal{D}$, G_{ij}	Algebraic terms defined in text
$\hbar$	Planck's constant $\div 2\pi$
$\underset{\sim}{H}$	Magnetic field
H_r	Harmonic average of r values
i	$\sqrt{-1}$
I	Intensity
j	Electric current density
J	Effective intensity
k	Absorption coefficient
$\underset{\sim}{k}$	Wave vector
$K(\lambda)$	Relative visibility of eye
L_a, L_c	Crystallite x-ray parameters
m	Mass of electron
n	Refractive index
$\underset{\sim}{\hat{n}}$	Director vector
N	Number of atoms per unit volume
$\hat{N}$	General refractive index
r	Reduced reflectivity
$\underset{\sim}{r}$	Position vector
r_i	(i = 1, 2) Radii of curvature of graphitic sheets
R	Reflectivity
$\mathcal{R}$	Percentage reflectivity
$\underset{\sim}{\hat{s}}$	Complex ray vector; direction of normal to a lamellar molecule
t	Time
T	A reflectivity parameter defined in the text

v	Speed of light in a material
α	Anisotropy ratio
Γ	Frequency related to width of absorption curve
δ_{ij}	Kronecker delta
ΔI	Intensity function
ε	Electric permittivity (dielectric constant)
θ	Phase change on reflection
$\hat{\theta}$	Angle between $\hat{n}$, $\underset{\sim}{z}$
$\hat{\theta}'$	Angle between $\hat{s}$, $\underset{\sim}{z}$
$\hat{\theta}''$	Angle between $\hat{s}$, $\underset{\sim}{\xi}$
$\hat{\theta}'''$	Angle between $\hat{n}$, $\underset{\sim}{\xi}$
θ_f, θ_s	Phase changes after passing through retarder plate
Θ	Orientation angle (rotation)
κ	Absorptive index
λ	Wavelength
μ	Magnetic permeability
ρ	Density; electric charge density; reflection coefficient
σ	Conductivity
ϕ, ϕ', ϕ''	General phase angles
Φ	Orientation angle
χ	Difference between θ_f and θ_s
ψ	Phase difference on reflection (model)
$\psi(\Phi)$	Phase difference on reflection (exact)
ω	Angular frequency; averaging angle
ω_o	Absorption frequency
Ω	Anisotropy index

ACKNOWLEDGMENTS

This study was supported financially be a Research Fellowship to C. C. from the British Carbonization Research Association and by a grant from the European Coal and Steel Community, Grant 7220-EB-807, to the Northern Carbon Research Laboratories. The authors express their appreciation to Miss B. A. Clow, Mrs. M. Poad, and Mrs. P. M. Wooster for care and patience in the preparation of the manuscript.

REFERENCES

1. S. Ergun, in *Chemistry and Physics of Carbon,* Vol. 3 (P. L. Walker, Jr., ed.), Marcel Dekker, New York, 1968, p. 45.

2. G. Davies, in *Chemistry and Physics of Carbon,* Vol. 13 (P. L. Walker, Jr., and P. A. Thrower, eds.), Marcel Dekker, New York, 1977, p. 1.

3. H. Marsh and C. Cornford, *ACS Symp. 21,* 266 (1976).

4. J. D. Brooks and G. H. Taylor, in *Chemistry and Physics of Carbon,* Vol. 4 (P. L. Walker, Jr., ed.), Marcel Dekker, New York, 1968, p. 243.

5. H. Marsh and P. L. Walker, Jr., in *Chemistry and Physics of Carbon,* Vol. 15 (P. L. Walker, Jr., and P. A. Thrower, eds.), Marcel Dekker, New York, 1979, p. 229.

6. E. Fitzer, K. Muller, and W. Schafer, in *Chemistry and Physics of Carbon,* Vol. 7 (P. L. Walker, Jr., ed.), Marcel Dekker, New York, 1971, p. 237.

7. A. Saupe, in *Liquid Crystals and Plastic Crystals* (G. W. Gray, ed.), Ellis Horwood, Chichester, West Sussex, England, 1974, p. 19.

8. P. G. de Gennes, *The Physics of Liquid Crystals,* Clarendon Press, Oxford, 1974.

9. J. D. Brooks and G. H. Taylor, *Carbon 3,* 185 (1965).

10. R. Didchenko, J. B. Barr, S. Chwastiak, I. C. Lewis, and L. S. Singer, *12th Biennial Conference on Carbon, Extended Abstracts,* American Carbon Society, 1975, p. 329.

11. J. L. White, *Prog. Solid State Chem. 9,* 59 (1975).

12. L. S. Singer and I. C. Lewis, *Carbon 16,* 417 (1978).

13. H. Marsh, *Fuel 52,* 205 (1973).

14. P. G. de Gennes, *The Physics of Liquid Crystals,* Clarendon Press, Oxford, 1974, p. 48.

15. Y. Yamada, T. Imamura, H. Kakiyama, H. Honda, S. Oi, and K. Fukuda, *Carbon 12,* 307 (1974).

16. R. T. Lewis, *12th Biennial Conference on Carbon, Extended Abstracts,* American Carbon Society, 1975, p. 215.

17. D. S. Hoover, A. Davis, A. J. Perotta, and W. Spackman, *14th Biennial Conference on Carbon, Extended Abstracts,* American Carbon Society, 1979, p. 393.

18. H. Marsh, F. Dachille, J. Melvin, and P. L. Walker, Jr., *Carbon 9,* 159 (1971).

19. G. H. Taylor, *Fuel 40,* 465 (1961).

20. R. A. Forrest and H. Marsh, *Proceedings of the Fifth International Conference on Industrial Carbon and Graphite,* Vol. 1, Society of Chemical Industry, London, 1978, p. 321.

21. R. A. Forrest, M.Sc. thesis, University of Newcastle upon Tyne, 1977.

22. D. L. Greenaway, G. Harbeke, F. Bassani, and E. Tosatti, *Phys. Rev. 178,* 1340 (1969).

23. M. Berman, H. R. Kerchner, and S. Ergun, *J. Opt. Soc. Am. 60,* 646 (1970).

24. E. S. Bomar and W. P. Eatherly, *Carbon 16,* 163 (1978).

25. M. Born and E. Wolf, *Principles of Optics*, 4th ed., Pergamon Press, Elmsford, N.Y., 1970.

26. L. P. Mosteller, Jr., and F. Wooten, *J. Opt. Soc. Am. 58,* 511 (1970).

27. L. D. Landau and E. M. Lifshitz, *Electrodynamics of Continuous Media,* Vol. 8 of *Course of Theoretical Physics,* Pergamon Press, Elmsford, N.Y., 1960.

28. F. Wooten, *Optical Properties of Solids,* Academic Press, New York, 1972.

29. J. L. White, *ACS Symp. 21,* 282 (1976).

30. J. L. White, *The Formation of Microstructure in Graphitizable Carbons,* Aerospace Report TR-0074 (4250-40)-1, 1974.

31. J. E. Zimmer and J. L. White, *Disclination Structures in Carbonaceous Mesophase and Graphite,* Aerospace Report, 1976.

32. J. Woodrow, B. W. Mott, and H. R. Haines, *Proc. Phys. Soc. B65,* 603 (1952).

33. H. Honda, H. Kimura, and Y. Sanada, *Carbon 9,* 695 (1971).

34. Y. Sanada, T. Furuta, H. Kimura, and H. Honda, *Carbon 10,* 644 (1972).

35. R. A. Forrest and H. Marsh, *Carbon 15,* 348 (1977).

36. R. C. Weast, ed., *Handbook of Chemistry and Physics, 53rd Edition,* Chemical Rubber Co., Cleveland, Ohio, 1972, p. E-209.

37. H. Kubota, in *Progress in Optics* (E. Wolf, ed.), Vol. 1, North-Holland, Amsterdam, 1961, p. 213.

38. L. L. Ban, *Chem. Soc. Lond. 1,* 54 (1972).

39. R. E. Franklin, *Acta. Crystallogr. 4,* 253 (1951).

40. E. Clar, *The Aromatic Sextet,* Wiley, New York, 1973.

41. J. T. McCartney and S. Ergun, *Fuel 37,* 272 (1958).

42. J. T. McCartney and S. Ergun, *U.S. Bur. Mines. Bull. 641.*

43. D. W. Van Krevelen, *Coal Science,* Elsevier, New York, 1961.

44. A. Davis, *Analytical Methods for Coal and Coal Products,* Vol. 1, Academic Press, New York, 1978, p. 27.

45. R. J. Marshall and D. J. Murchison, *Fuel 50,* 4 (1971).

46. D. Chandra and R. L. Bond, *Proceedings of the International Commission on Coal Petrology,* Leige, 1955, No. 2, p. 47.

47. R. L. Bond and D. Chandra, *The Reflectance of Carbonized Coals,* BCURA Information Circular 139.

48. D. Chandra, *Fuel 44,* 171 (1965).

49. D. Chandra, *Econ. Geol. 60,* 621 (1965).

50. A. Davis, Ph.D. thesis, University of Newcastle upon Tyne, 1965.

51. T. K. Ghosh, *Econ. Geol. 63,* 182 (1968).

52. H. A. W. de Vries, P. J. Habets, and C. Bokoven, *Brenst. Chem. 49,* 105 (1968).

53. F. Goodarzi and D. G. Murchison, *Fuel 51,* 322 (1972).

54. F. Goodarzi and D. G. Murchison, *Fuel 52,* 164 (1973).

55. J. W. Patrick, *The Influence of Carbonization Conditions on the Development of Optical Anisotropy in Coke,* Report of British Carbonization Research Association, Chesterfield, England.

56. K. J. Huttinger and U. Rosenblatt, *Abstracts of the Fourth International Conference on Industrial Carbon and Graphite,* Society of Chemical Industry, London, 1974, p. 8.

57. H. Piller and K. Van Gehler, *Am. Mineral. 49,* 867 (1964).

58. H. Marsh, J. W. Akitt, J. M. Hurley, J. Melvin, and A. P. Warburton, *J. Appl. Chem. 21,* 251 (1971).

59. I. Macefield, M.Sc. thesis, University of Newcastle upon Tyne, 1977.

60. R. A. Forrest, M. French, H. Marsh, J. A. Griffiths, and J. L. White, *13th Conference on Carbon, Extended Abstracts,* American Carbon Society, 1977, p. 207.

61. H. Marsh, unpublished results, 1980.

62. J. D. Jackson, *Classical Electrodynamics,* 2nd ed., Wiley, New York, 1975, p. 241.

Author Index

Numbers in parentheses are reference numbers and indicate that an author's work is referred to in the text. Underlined numbers give the page on which the complete reference is listed.

Abajev, S. S., 84(28), <u>159</u>
Akitt, J. W., 303(58), <u>330</u>
Albert, P., 14(20), <u>62</u>
Altshuler, B. N., 84(27), 134 (27), <u>159</u>, <u>160</u>
Arefieva, E. F., 76(11), 79(11), 86(33), 96(38,39), 101(38), 131(49), 133(49), <u>158</u>, <u>159</u>, <u>160</u>
Austin, L. G., 176(20), 183(28), 201(20), 206(20), <u>209</u>, <u>210</u>

Baird, T., 147(61,63), <u>160</u>
Baker, R. T. K., 165(4), 195(4, 37,38), <u>209</u>, <u>210</u>
Ban, L. L., 265(38), <u>329</u>
Barr, J. B., 217(10), <u>328</u>
Bassani, F., 223(22), <u>329</u>
Berman, M., 223(23), 242(23), <u>329</u>
Biederman, D. L., 201(45), 202 (45), <u>210</u>
Bodenstein, P., 77(13,14), <u>158</u>
Bokoven, C., 277(52), 298(52), <u>330</u>

Bokros, J. C., 42(57), <u>63</u>, 66(2), 69(2), 96(2), <u>157</u>
Bomar, E. S., 223(24), 240(24), <u>329</u>
Bond, R. L., 277(46,47), <u>330</u>
Born, M., 223(25), <u>329</u>
Borodina, L. M., 76(10), 79(10,20, 21,26), 85(20,26), 86(34), 94 (35), 101(10), 132(10), 146(26), 147(20,26), <u>158</u>, <u>159</u>
Boudart, M., 198(42), <u>210</u>
Bowman, J. C., 7(10), <u>61</u>
Brooks, J. D., 213(4), 214(4), 215 (4), 217(9), 219(4), 221(4,9), 298(9), <u>328</u>

Castang-Coutou, M. F., 23(37), <u>63</u>
Cen, P. L., 172(17), 189(33), 191 (33), 195(33), 199(43), 205(47), 206(50), <u>209</u>
Cerović, D., 29(48,49), 30(48), 31 (48), 32(48), 52(67,68), 54(67), 55(67), 56(67,68), 57(68), 58(67, 68), <u>63</u>, <u>64</u>
Chandra, D., 277(46,47,48,49), <u>330</u>

Subject Index

Printed and bound by CPI Group (UK) Ltd, Croydon, CR0 4YY

17/10/2024

01775696-0014